Water, Earth, and Sky

The Colorado River Basin

Water, Earth, and Sky
The Colorado River Basin

Photographs by Michael Collier

Foreword by David L. Wegner

Essays by

Michael Collier

John C. Schmidt

E. D. Andrews

Richard A. Valdez

Lawrence E. Stevens

Ellen Meloy

The University of Utah Press
Salt Lake City

"The Silk That Hurls Us Down Its Spine," copyright © 1998 by Ellen Meloy, is printed courtesy of the author.

Printed in Hong Kong

        Library of Congress Cataloging-in-Publication Data
Water, earth, and sky : the Colorado River Basin / photographs by
  Michael Collier and foreword by David L. Wegner : essays by Michael
  Collier ... [et al.].
        p.   cm.
   ISBN 0-87480-598-8
   1. Colorado River Watershed (Colo.-Mexico) I. Collier, Michael.
GB122.7.C6W38   1999
551.48'2'097913—dc21   98-41925

Frontispiece: Green River at the head of Labyrinth Canyon, Utah.
Above: Sycamore leaves.
Facing page: Reflections in Grand Canyon.

Contents

A NOTE TO THE READER ABOUT THIS BOOK. The photographs constitute the basic "text" of this book and depict various features of the Colorado River system as seen from the air. They essentially trace the river system from top to bottom, or start to finish, as it were. Because the river system is huge and complex, however, certain choices of arrangement had to be made. First, the mountains that constitute the beginnings of the system are illustrated at the start without distinction as to their location in Wyoming, Colorado, or Utah. The Green River is then traced from its headwaters in Wyoming to its end—the confluence with the Colorado at Canyonlands National Park in Utah. Backtracking somewhat, the Colorado (formerly Grand) River is traced next from its highlands in Colorado to the confluence with the Green; the Colorado River is then followed on its course to the ocean, with various tributaries picked up (and sometimes traced back to their sources) as they intersect the main river. The individual essays that accompany the photographs are not to be seen as a comprehensive treatment of the river system but as individual insights into aspects of the whole, to be browsed and/or studied at leisure. The few larger photographs at the beginning and end of the book are not part of the continuous photographic sequence. They illustrate areas already treated, giving another view of the Grand Canyon or the mouth of the river at the Sea of Cortez, for example.

ACKNOWLEDGMENTS

THE COLORADO RIVER BASIN IS BLESSED with many good people who love to call this land their home. The authors wish to thank the following individuals and institutions for their support and encouragement of this project: Ginger and John Giovale; Rob Elliott of Arizona Raft Adventures; David Houchin and Jay Krueger of Arizona Air-Craftsman; Steven Carothers of SWCA Environmental Consultants; Edie Crawford and the Schniewinds of Canyon Explorations; Mark Thatcher of Teva Sport Sandal; Jack Coles of Nature's Scene.

This book sprang to life during conversations with Dave Wegner, then the director of the Glen Canyon Environmental Studies unit of the U.S. Bureau of Reclamation and now the vice president of the Glen Canyon Institute. His efforts toward understanding and sustaining the Colorado River have already and will continue to make a difference.

Dawn Marano at the University of Utah Press has been all the editor that any writer could ask; without her, this book would not have seen the light of day. Many thanks to the Water Education Foundation in Sacramento for use of their map of the Colorado River Basin.

I think back fondly of the time I spent on the Colorado with David Lowry, Wesley Smith, David Edwards, Louise Teal, Drifter Smith, Lorna Corson, George Ruffner, and all the other wonderful people of the river. Rowing with them, I learned that every stroke of the oars is a prayer. Bob Webb, my partner in bureaucratic crimes and imaginary misdemeanors at the U.S. Geological Survey, continues to help make adventures like this one happen. And Rose Houk makes it all worth it.

Michael Collier
Flagstaff, Arizona
March 1998

Facing page: The muddy Little Colorado River at its confluence with the Colorado River, Grand Canyon, Arizona.

Above: Thunder River, Arizona.

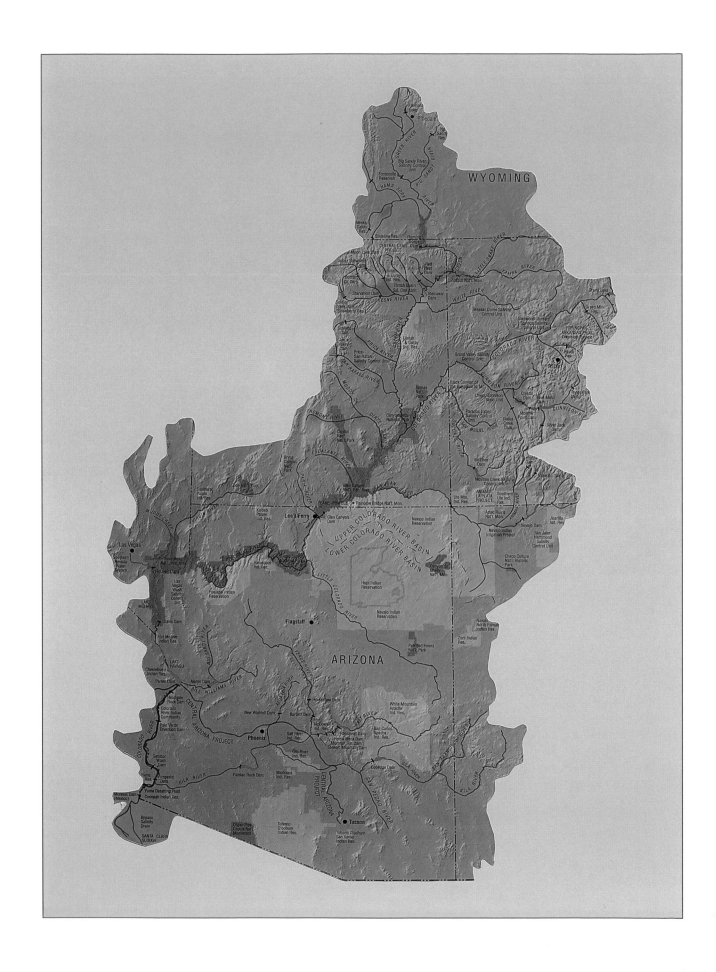

A Watershed Shaped by Time

David L. Wegner

Facing page: Map of Colorado River system.

Above: Vase flower.

THE BOOK YOU HOLD began as a conversation that I had with Michael Collier in 1995. I was interested in exploring landscape-level scientific approaches to evaluating riparian and aquatic habitats of the Colorado River system. I was looking for an approach that would transcend the historical and conventional tendency to fragment ecosystems such as the Colorado River Basin into meaningless micro-habitats. I wanted a natural history perspective that would integrate all aspects of this landscape. I wanted to present a perspective that appreciated this landscape as a continuum of habitats, a watershed shaped by the intricate interaction of many factors over time: geologic time, hydrologic time, biologic time. It was clear to me that Michael's stunning aerial photographs of the basin supported that vision, his images providing not only fine art, but a compelling visual pedagogy and a map of what the Colorado River system comprises.

The story that this book illustrates began more than 500 million years ago as the first fingers of the ancient oceans moved inland and began to lay down the sediments that today characterize the canyon walls throughout the basin. These sediments and those reshaped by the hot desert winds are recognized by dozens of vibrant names—Mancos Shale, Navajo Sandstone, Kaibab Limestone, and Moenkopi Shale—and evoke an image of an emerging landscape defined by a long, rich history of repeated deposition, aeolian reshaping, and resulting erosion. These seas brought with them the origins of life and, as they receded, left landforms that would ultimately become the mesas and canyons, the valleys and floodplains around, through, and over which the modern Colorado River flows.

It is a troubling exercise, to say the least, to juxtapose these imaginings of the nascent Colorado River Basin and the natural processes that spanned ten or twenty million years and shaped its ecosystems against what our civilization commonly defines as progress.

Facing page: La Plata Mountains northwest of Durango, Colorado, among the mountains where the Colorado River has its beginning.

Above: Monument Valley, Arizona/Utah, one of the most spectacular of the eroded landscapes of the Colorado River system.

The Colorado River delta, proclaimed by Aldo Leopold to be one of the most biologically rich and diverse ecosystems on the planet just a scant eighty years ago, is today a hollowed out and desiccated remnant. By the end of 1997 less than ten percent of the original wetlands were left. Those vestigial wetlands are supported only by salty irrigation return flows—a far cry from what the system once was like. The annual spring floods that were necessary to rejuvenate and maintain the biological richness of the delta are now cut off by forty-six major dams or diverted into agricultural fields growing subsidized crops. The dams have fragmented the river system, and the returning runoff from cities and agricultural fields make the water in many cases unfit for consumptive use. Many of the unique assemblages of native species of plants and animals that evolved in relative isolation over millions of years have been unable to adapt to rapid changes caused by humans and are close to extinction. Artificially created reservoirs have drowned hundreds upon thousands of Native American artifacts and with them the cultural heritage of the ancient peoples who lived in the basin.

Yet despite the assaults that are taking the life of this magnificent

river, there is still some cause for hope. People today *are* awakening to the fact that cheap water is expensive and realizing that the price for golf courses, fountains, and swimming pools in a desert is much higher than they thought. The cost for these amenities can be measured in impacted resources, in drowned canyons, in exterminated species, and in lost opportunities for future generations to gaze in wonder at the canyons and the river. What we don't yet know is whether our efforts to mitigate the damage already done to the Colorado and to retool (or abandon) our concept of progress, will come too late. I have hope and a passionate belief that we as humans can revisit and rethink the logic of the past and find ways to protect and restore many of these priceless resources. But as this book is going to press, more than two hundred new toilets are still being installed in Las Vegas alone—every day.

*

The Colorado River and its basin have been a crucible of evolution, a pathway for colonization, a hothouse for rampant development, a political and social cauldron, and a spiritual sanctuary. The six contributors whose essays accompany Michael's photographs explore the history of this watershed, both natural and human. They do not provide an exhaustive body of reference information, but rather an inviting interpretive context for the visual images. In the same way that Michael may change the setting on the aperture of his camera, opening or closing it to admit varying amounts of light, these essays illuminate chosen aspects of this landscape: the life cycle of a razorback sucker; the conditions that shape the meanders of a river; the sensory feast of rock, air, water, leaf, and bird that may nourish the human soul.

Geology, hydrology, biology, ecology, photographic and literary art: for the purposes of this book, these are not rigid disciplines so much as they are lenses through which the contributors to this book help us enter understanding. Michael Collier, trained as a geologist, has been running rivers, flying, and photographing the Southwest for twenty-five years. Jack Schmidt is a geomorphologist who has developed fundamental concepts of how river banks respond to moving water. Ned Andrews is a U.S. Geological Survey research hydrologist who has spent decades investigating how the Colorado and other rivers flow through and affect a myriad of landscapes. Rich Valdez is acknowledged as one of the country's foremost experts on fish within rivers of the mountain West. Larry Stevens is a research biologist who has spent twenty years inside Grand Canyon researching its riparian ecology. And Ellen Meloy is a prize-winning writer of natural history who has been on hundreds of trips down rivers of the Colorado Plateau, alone and accompanied by her husband, a river ranger in Utah.

Water, Earth, and Sky: The Colorado River Basin is an invitation to an environment of wonder, a landscape of imagination. For centuries people have stood on the edges of its mountains, canyons, and rivers peering in, awestruck and humbled. This book is, as Michael and I hoped it would be, a mosaic of splendid parts, a unique and dynamic assemblage like the Colorado River Basin itself, which, viewed from the right distance, resolves into the patterns that make a marvelous and inspiring whole.

Facing page: Tidal flats, mouth of Colorado River at the Sea of Cortez, Sonora/Baja California, Mexico.

Water, Earth, and Sky: The Colorado River Basin

Mags Hot, Master On: Flying For The Vision

Michael Collier

Facing page: San Francisco Peaks north of Flagstaff, Arizona.
Clouds, earth, and sky as seen from the air above a part of the Colorado River Basin.

Above: Brittlebush.

A FEW YEARS BACK I flew two friends from Flagstaff to Lees Ferry. One sat up front, enjoying the flight and talking too much. The San Francisco Peaks, S.P. Crater, and Gray Mountain streamed by. I went into a daze, absently highlighting my passenger's monologue with "hmmm's" and "uh-huh's." The air was as smooth as the glassy lakes I once canoed at first light in my youth. We skimmed over the Marble Platform, five miles east of the Colorado River. Morning light washed up on Shinumo Altar, and poured like liquid gold over the rim into Marble Canyon. We glided above brown Navajo hogans and empty sheep pens. Every arroyo was etched in perfect detail as it tumbled toward the river.

My friend Fran Joseph sat in the back seat. This country is in her blood. She has hiked and boated here for decades, but had never seen it from this perspective. Fran hadn't said a word over the intercom since we departed. I turned around to make sure she was okay. Tears were streaming down her face. "It's so beautiful," she said with a smile that looked like sunbeams between shafts of rain. "It is just *so* beautiful."

I love to fly. Since boyhood, I have been blessed with flying dreams: moments of perfect happiness, crystallized into the sudden ability to wish myself aloft. These dreams now come true on the wings of my 1955 Cessna. I fly for any number of reasons—to get from one place to another, to explore geologic features, to photograph. I fly for the view, but mostly I fly for the vision.

A small airplane is a fine platform from which to learn a landscape, a thousand feet above ground. From above, the earth becomes a mosaic of ridges, valleys, and plains. I see the obvious: rivers flow from the mountains down to the sea. And I see the not-so-obvious as well. My mentor Chris Condit once kept track of our progress from Flagstaff to Santa Fe by reading the highway signs along Interstate 40. As we passed Gallup, snow swirled into the cockpit around the loose edges of his airplane's

Facing page, top: Glacial ice in the Wind River Mountains near Gannett Peak, Wyoming.

Facing page, bottom: Collegiate Range of the Rocky Mountains near Buena Vista, Colorado. Water within the Colorado River system comes predominantly from the basin's high country— the Rocky Mountains and San Juans in Colorado, the Wind River Mountains in Wyoming, and the Uintas in Utah.

Above: Headwaters of the South Fork of the Little Wind River, in the Wind River Mountains, Wyoming.
The Wind River Mountains form the Continental Divide in western Wyoming, with meltwaters flowing east into the Platte River and west into the Green River.

windshield. I was a student then, learning bad habits and loving every minute of it. More recently, my wife and I approached Gallup on a crystal-clear morning, skirting the edge of a dying thunderstorm. A hard rain had fallen overnight. A great cumulus cloud was gliding east; we were headed west.

Over the Painted Desert, we curled around fluffy patches of fog that hugged the ground. Light seeped through the clouds, reflecting off a thousand pools of rain water scattered among the green, brown, and red shales. Past Holbrook, I saw mud parked high above the banks of Leroux Wash and realized that we had just missed seeing a good-sized flash flood. But fifteen miles farther west, the Little Colorado River was in full flood— five thousand very brown cubic feet per second rushing downstream, madly mixing foam, logs, old tires, and tumbleweeds. Then over Winslow, we saw that the Little Colorado's bed was still dry. The flood had not yet arrived. Thunderstorm, fog, pools, and flood. The airplane had shown us the entire life span of the storm compressed into a few minutes, as if we were watching a film run in reverse at 150 miles an hour.

An overflight can never replace physical encounters with a landscape, of course. There is no substitute for the immediate sensations that fill even the shortest walk back down on Earth: the smell of sand verbena, the murmur of running water, the chill of rain on one's face, the chafe of sandstone, the diffraction of sunlight through a dewdrop on lupine leaves. The aerial perspective is necessarily an overview, where individual vignettes learned at ground level are woven into a larger story of the land. From above, restricted to neither roads nor rails, I have discovered how much has been disturbed in some places, and how little in others. From the cockpit, I have seen how the earth fits together.

*

The Colorado River Basin sprawls across 244,000 square miles of Wyoming, Colorado, Utah, New Mexico, Arizona, Nevada, and California, and borders the Mexican states of Sonora and Baja California. Thirty-five perennial rivers and innumerable ephemeral streams all contribute what they can to the mainstem Colorado: water here, sediment there, and a dash of salt along the way. When all is said and done—after twenty million people have quenched their thirsts, after three million acres have been irrigated, after 11.5 billion kilowatt-hours have been generated beneath dams whose lakes evaporate well over a million acre-feet of water each year, after one-third of the river's flow is diverted to places like Denver, Salt Lake, and Los Angeles—no water is left when the Colorado's dry bed meets the Sea of Cortez.

The Colorado is small compared to other rivers in the United States, carrying eight percent of the volume of the Columbia River, and only three percent of the Mississippi. Tree-ring studies indicate that for the last few hundred years the average annual flow of the Colorado has been 13.7 million acre-feet (each of which would cover an acre of land under one foot of water). Some recent ten-year running averages have been as high as 18.8 million acre-feet per year, others as low as seven million.

Five-sixths of this water originates in the mountains of Colorado and Wyoming, an area covering less than fifteen percent of the basin. Undammed, seventy percent of the Colorado's water would flow downstream during only three months of the year—May, June, and July.

Ecologist Larry Stevens figures that every drop of the Colorado River is used and re-used seventeen times before finally being evaporated, diverted, or—in a rare year of heavy flow—allowed to reach the Sea of Cortez. With so much being asked of this relatively small river, it's no surprise that we become entangled in the politics of water allotments, in dogfights over dam operations, in squabbles about how much salinity who is contributing where along the river. It's no surprise that we lose sight of the river itself.

After twenty-five years of river-running, after rowing and paddling eleven thousand miles on the Colorado and its tributaries, I wondered what the river and its basin really looked like. In 1996, I set out in my Cessna, seeking an understanding of the river and its basin as they exist in the real world. I shrugged off political boundaries, set aside academic arguments, and perhaps even ignored an occasional FAA regulation. The photographs in this book are the result of that search.

What I found is a river that has been changed, but remains a river all the same. Below spreads a vibrant landscape with much more living habitat than I had imagined. At fourteen thousand feet, I rolled wildly amidst the Wind River and Neversummer Mountains, tracing thin lines of the nascent Green and Colorado Rivers through talus and tundra at their headwaters. I followed waves of Chinle Shale flushed down the Dead River into the Puerco, the Little Colorado, and on through Grand Canyon. I watched the river go limp and then lifeless as the last of its water was siphoned into fields near San Luis, Sonora.

Vibrant here, lifeless there. Is it possible to embrace both ends of this paradox? I believe so. The first step is to temporarily suspend the political and cultural filters that so often color our vision of the world. I squint down and see water, a bed, banks. Sandbars, cutbanks, and current patterns. Cottonwoods, sedges, and sycamores. I look at the processes at work: sediment aggrading here, degrading there. Banks retreating in places, or migrating into a narrowing channel elsewhere. Now I integrate this view of the river over time: a memory of the Green River cresting last spring, a print-out of the stream-gauge measurements at Lees Ferry stretching back to 1895, a history of steamboats churning up the Colorado past Yuma and Ehrenberg.

Only after a thorough and honest assessment of the river at a given location am I ready to consider the forces acting upstream and down. Yes, the Parker Strip on the Arizona-California border is a little bit of hell on earth. Yes, Las Vegas would just as soon suck the Virgin River completely dry below Zion. But even so, I don't want to lose sight of the river as it is—right here, right now. The Colorado may be channelized between dikes at Needles, but that doesn't undermine my sense of wonder as I watch a flock of snow geese splash into a back bay beneath the Trigo Mountains. I may long with all my soul to revisit the sandstone cathedrals drowned beneath Lake Powell, but that doesn't mean that I am unable to soak in a half-billion years of solitude when I sit within the

Facing page: Maroon Creek in the Elk Mountains above Aspen, Colorado.

Above: Mount Sneffels above Telluride, Colorado, in the San Juan Mountains.

Right: Maroon Bells of the Rocky Mountains near Aspen, Colorado.

sculptured confines of Blacktail Canyon. The Colorado may not flow at all below San Luis, but that does not diminish the awesome power of the river I have felt when rowing Lava Falls in Grand Canyon.

*

From the air I see geologic lessons spoken in the vocabulary of uplift and erosion. The Colorado River Basin is an amalgam of mountains and valleys, high plains and deserts, all stitched together by flowing water. The upper Crystal River wears away at the Paleozoic backsides of the Maroon Bells, descending through ice-choked glacial ponds and impossibly beautiful hanging valleys. The Gunnison cleaves Black Canyon, slicing through the rising metamorphic basement rocks like a hot wire through ice. The San Rafael uncoils itself from the sedimentary Sinbad Country, a snake made of more sand than water. The Hassayampa River joins the Gila on their way to the Colorado, a combined length of 150 miles along a channel that is bone dry far more often than not. From the air, the seams between the peaks, valleys, deserts and canyons seem smoother. From the air, it becomes obvious that the elements of this landscape are interconnected and that basins are best considered as a whole.

Flying downstream, I watch the Yampa mix with the Green, then join the Colorado. The creamy Paria and brown Little Colorado Rivers arrive near the head of Grand Canyon. With each confluence, the Colorado swells a bit more, taking on a little of the character of each new tributary—steady or flashy, clear or turbid. The Colorado shrieks through Gore Canyon, lolls past cottonwoods beneath Battlement Mesa, and

Above: Cottonwood Creek in Soap Hole Basin north of Big Piney, Wyoming.

Above: Farms along the Green River downstream from Daniel, Wyoming.

Right: Meanders on the Green River at Big Twin Creek upstream from Bronx, Wyoming.
The Green River periodically adjusts its course into meanders according to the variables of gradient, volume, sediment, and channel roughness. The meanders can be cut off when the balance of these variables changes, such as during a large spring flood.

Above: Green River at Bronx, Wyoming. Ranching has been an integral aspect of the Green River landscape since settlement days in the second half of the nineteenth century.

rumbles through the Big Drops of Cataract Canyon. Memories of the San Juan and Escalante silently mingle with the mainstem beneath the surface of Lake Powell. The Green, Gunnison, and Colorado Rivers sleepwalk through reservoirs at Flaming Gorge, Curecanti, and Lake Mohave. Below the dams, the rivers awaken and try to remember what and where they are.

Aloft, I can only imagine the life that I know teems within the river. Over Desolation Canyon, I look down on the slough where I helped Jack Schmidt and Rich Valdez seine for native flannelmouth suckers and bonytail chubs. These backwaters are essential to young fish, native or not. Here the current slackens and the water temperature rises. The morphology of these backwaters, or secondary channels, is determined by floods and flows, sediment and stabilization. From the air, I see how sandbars isolate the backwaters and fit into the overall circulation of the river.

Most striking of all is the irrepressibility of riparian habitat. Willows line the meanders of Cottonwood Creek in Wyoming's Soap Hole Basin.

Douglas firs cloak the banks of the Dolores River. Stately sycamores arch
over Aravaipa Creek at the north end of the Galiuro Mountains. An im-
penetrable mesquite and tamarisk thicket clogs the river's delta in Mexico.
Because of changes in streamside habitat, bald eagles and peregrine fal-
cons now thrive in the Grand Canyon, and endangered willow flycatchers have found a foothold amidst tamarisks at the head of Lake Mead.

All landscapes change, some faster than others. Rocks spall into steep
glacial valleys; debris-flows race down canyon washes; wind hefts tons of
desert dust before an oncoming storm. Landscapes evolve fitfully in some
places, more steadily in others. Seen from the air and from a geologic
perspective, the Little Colorado River's course near Cameron has repeat-
edly been rearranged by lava pouring into its channel from the south.
Innumerable embayments line the Colorado's course through the Imperial National Wildlife Refuge—dim outlines of a 1930s reservoir that the
persistent, sediment-choked river is gradually filling and reclaiming.

In 1990 Robert Webb of the U.S. Geological Survey reshot Robert
Brewster Stanton's 1890 photographs of the Grand Canyon. The matched
photos give testament to the incredible variability of change in the canyon. It is possible to find precarious flakes of rock still in place that should
have blown away a century ago. The photos document individuals from
forty-one plant species that have survived at least a hundred years. But
the photographs also bear witness to sudden catastrophic change—sometimes natural, sometimes of human origin. Debris-flows have swept down
tributaries in Grand Canyon on average every twenty-five years, instantly
extending alluvial fans far out into the river. Tamarisks have invaded many
stretches; willows now thrive where once they would have been washed
away by spring floods. Sand has been stripped at a feverish pace from
many beaches and bars in Grand Canyon since Glen Canyon Dam was
built in 1963.

To be sure, habitat changes can be ecologically unsettling. Venerable

Above: Green River at Jensen, Utah.
Old meanders within the floodplain are thickly vegetated and become critical habitat for animals living along the river.

Right: Green River at Horseshoe Bend south of Vernal, Utah. At high flows, the Green River splits into multiple channels and then overtops and rearranges its sandbars. Once the water drops, non-dominant channels become warm backwaters where native fish spawn. If not disrupted by periodic flooding, vegetation will stabilize these sandbars and the backwaters will eventually disappear.

riverman Martin Litton, longing for the days of an unfettered river, grumps, "How many willow flycatchers do we need?" But ecologist Steve Carothers shrugs, asking why should we expect that a changing river will remain static, just to preserve what we *think* it looked like a century ago. I gaze down from my plane on these arguments, on this river, and once again I see stretches that are vibrant, stretches that are lifeless. The hand of man has undeniably stirred this soup. And still the river flows from the mountains to the sea.

It would be foolish—indeed impossible—to ignore human impact within the river system. Levees straighten and sterilize much of the Colorado where it flows along the border between Arizona and California. There are at least 122 formidable dams within the basin, forty-six of them able to significantly impede the flow of its biggest rivers. Beyond this, it is impossible to accurately estimate the number of smaller water-retention structures bulldozed across nearly every drainage within the basin. As hydrologist Ned Andrews explains, evaporation losses long ago outweighed the reservoirs' combined storage benefits. Populations within and around the basin have skyrocketed. California consistently withdraws more than its allotted 4.4 million acre-feet of river water every year.

But at the same time, a living river still flows through this basin. It runs silently through the reservoirs, it runs low after the diversions, but by god it runs. Great blue herons come home at sunset to roost below DeBeque. River otters play in the lower reaches of Black Canyon of the Gunnison. Schools of Sonoran suckers and roundtail chubs patrol pools of the Verde River below Perkinsville. Cottonwood galleries line the Uncompahgre River below Ouray. The Salt River can still mount a 143,000-cubic-foot-per-second flood to ream out its channel and deposit fresh sand forty feet above the water's summertime surface.

Ellen Meloy, writing about the Green River through Desolation Canyon, got it right: "I am still amazed that we call this organism 'the River' when it is actually several rivers of the Colorado River system, numerous segments with no resemblance to a river, and a few more with no water in them whatsoever." It's not enough to consider single rivers or isolated stretches of river. To understand a landscape, think in terms of its interlinked drainages. A river basin is the basic building block of geomorphology—which is why nineteenth-century explorer John Wesley Powell advocated laying out the West, not by the surveyor's straight line, but along the twisting boundary of a basin divide. Powell argued for treating river basins as if their water and sediment moved coherently downstream. Fortunately scientists and some land managers are now validating and advancing this sound seminal thinking.

When the U.S. Bureau of Reclamation began to divert water from the upper Green River through the Central Utah Project to the west slope of the Wasatch Mountains, the U.S. Fish and Wildlife Service protested. Native humpback chubs and Colorado squawfish, it argued, were already up against the ropes of extinction—a situation that the diversion would only exacerbate. In 1980, the Fish and Wildlife Service imposed a Biological Opinion demanding a Reasonable and Prudent Alternative. The Central Utah Project could proceed, but Flaming Gorge Dam above Dinosaur National Monument would have to be operated to approximate

Left: Birds on a Green River sandbar near Ouray, Utah.

Below: Dead cottonwood trees along the Green River at Ouray, Utah.
The cottonwood is dependent throughout its life-cycle on floods. Its seeds require flooding for dispersal and germination; its shallow roots need the river's nearby moisture. These cottonwoods were killed by the highwater floods of 1983.

Above: White River just above Ouray, Utah. Sand is deposited on the insides of meanders where the current is slow. Vegetation crowds the floodplain and will eventually stabilize the sandbar in the middle of the photograph.

natural flows in which the native fish had evolved. In addition the dam would have to regulate the Green River in conjunction with flows on the undammed Yampa a few miles downstream, so that spring floods would once again coincide with spring spawning, so that high summer flows would fall to steady wintertime lows. In an effort to preserve habitat for native fish, the Bureau of Reclamation was forced to trade water withdrawals for the opportunity to generate high-priced peaking power at Flaming Gorge Dam.

Before Glen Canyon Dam was built, the Colorado River rose to 86,000 cubic feet per second in Grand Canyon most every spring. In March 1996, an intentional flood was released from the dam specifically designed to mimic the pre-dam annual flood and thus to improve downstream conditions in Grand Canyon. Never before had environmental concerns held equal footing with the other mandates by which this dam is operated: flood control, water storage, power generation. What did the four million dollars spent on the flood buy?: a few expected results, a few

surprises, and a good deal of information. Beaches were rejuvenated for a while. Some alluvial fans were beveled down, making their rapids more navigable for a time. The flood was not big enough to significantly affect trees and shrubs within the Canyon's riparian corridor. Archaeological sites were neither destroyed nor protected by the shifting of sediment. And the fish—both native and non-native—seemed to have just moved out of the way and watched the flood go by, no doubt interested but unaffected.

Ultimately though, the most important benefit of the flood was the sense that it is worth trying to manage a river as if quality, not just quantity, of flow really mattered. Since at least the Pleistocene, this river and its riparian life have evolved with the anticipation that most of an entire year's flow would come down in springtime. The experiment was an acknowledgment that floods are integral to the nature of Southwestern rivers.

Left: Green River just below Sand Wash, Utah.

Above: Maples northwest of Standardville, Utah.

The 1996 flood focused the world's attention on the intelligent management of dams and rivers for a few weeks. Can this attention last? Can these ideas be extrapolated into the future and expanded to include an entire basin? Laguna Dam, built in 1909 above Yuma, was the first permanent structure ever to harness the Colorado. In the subsequent sixty years, dams and diversions blossomed up and down the river's length. Then in the 1970s the age of dam building ground to a halt. We now must live with what we have wrought. Yet having so far survived this century of technological depredation, we must not compound our errors by failing to appreciate and protect what is left—*all* of what is left, not just a few national parks and wildlife refuges, and a fistful of endangered species. We must see the basin for what it is, here and now. Only with this clarity of vision can we begin to make intelligent decisions about what we think the river *should* look like in the future.

*

Mags hot. Master on. Full rich. "CLEAR PROP!" I yell, and punch the starter of my forty-year-old plane. Six cylinders, each the size of a coffee can, pop, sputter, and then roar to life. I taxi downwind with a loping gait on tires that aren't quite round, checking the controls, the radios, the engine. I line up on the runway and ease in full throttle. The Cessna gathers speed, the tail lifts and we're airborne! Always this moment is pure magic. I soar as the plane climbs.

I circle up and out of Glenwood, following the Roaring Fork as it joins the Colorado. Late afternoon sunlight streams into the cockpit as we turn west, heading downstream. Interstate 70 and the Denver and Rio Grande tracks bracket the river left and right. At a thousand feet, I'm dwarfed by the Roan Cliffs towering overhead. I look for herons in the cottonwoods as we pass DeBeque, and cut behind the Book Cliffs to bypass Grand Junction.

Ruby Canyon. Twentyeight Hole Wash. Westwater Canyon. The cliffs rise and fall, breathe in and again out. Small irrigated farms punctuate the bottomlands. The Dolores rolls in from the San Juan Mountains. Onion Creek. Professor Creek. Salt Wash. I give the plane some rein; she knows where the loose rocks are, where the prairie dog holes are. There's no sense in having too tight a grip; after all, she started flying twenty years before I ever did.

On downstream, past the potash ponds and into Canyonlands. I wheel and turn with a raven over Upheaval Dome, rising on a thermal and slicing sideways back down. The wings are rarely level but the engine sounds good, gliding over the Green within Stillwater Canyon, floating past the Maze, heading toward the Orange Cliffs. I photograph, intoxicated by the light and by the land as the sky slips from blue through gold to crimson. The plane curls through the sky, drawing circles tangent to the lines I see in the viewfinder. As the sun sets and the alpenglow dims, I set my camera aside and gather my bearings. Not lost, but living in a different world. I'm guided by the inevitability of this river heading for the sea, through-flowing, alive, and magnificent.

Left: Price River as it enters the Book Cliffs near Woodside, Utah.
The Price River flows through shales that shed a tremendous quantity of fine sediment with each passing rainstorm. The river anomalously dives into the Book Cliffs instead of going around them on its way to join the Green River below Desolation Canyon.

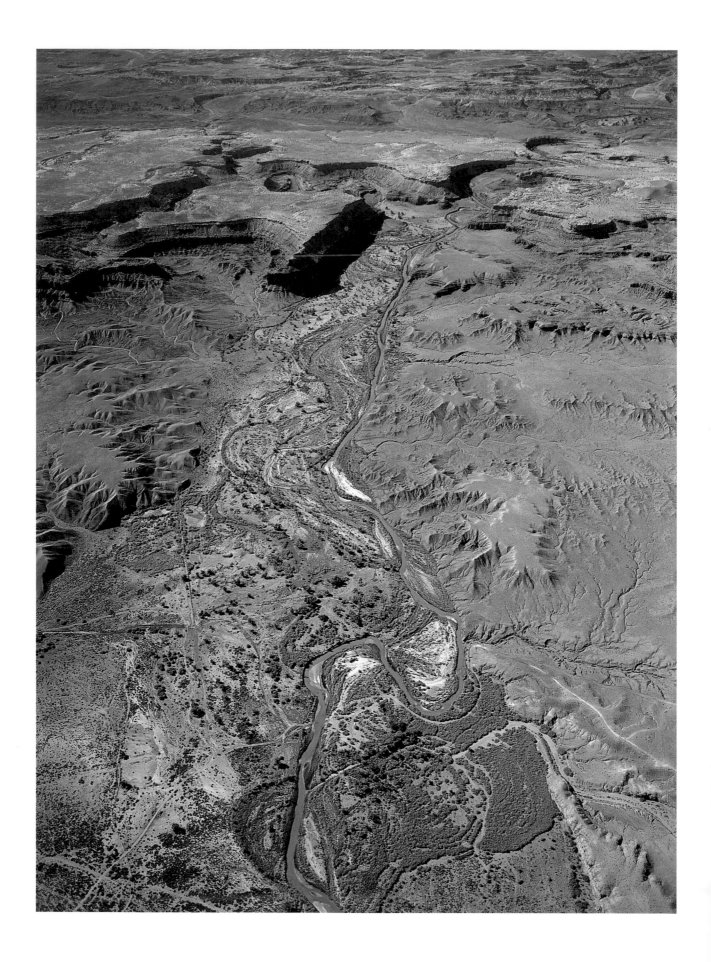

Rock-shaped River in River-shaped Rock

John C. Schmidt

Facing page: Huntington Creek in Castle Valley, Utah.
Like so many of the Green River's tributaries in Utah, Huntington Creek adds proportionately far more sediment than water to the mainstem river.

Above: Grand Canyon sand deposits.

EVERY BOATMAN WHO STANDS surveying a set of rapids from a bank of the Colorado reads that churning mix of white water and mud as a thrilling collaborative story written by river and by land. There: the slow water above the constriction. There: the constriction itself—partially submerged boulders, some of them house-sized, dumped into the channel from a side canyon during a massive debris-flow. And there: the eddy—the pocket of calm water below every debris fan—doubling back to a large sand beach where the grateful oarsman and his passengers might pull over for lunch.

In the Colorado River this trio of distinct features, slow water, rapid, and eddy, occurs again and again—more than 200 times in the Grand Canyon alone, and nearly as frequently in the canyons of Westwater, Cataract, Desolation/Gray, and Dinosaur National Monument. We could call this the story of the rapids, and there are many others, just as distinct and memorable: the story of the delta that the Colorado and the desert tell at the Sea of Cortez; the story of strata, of geologic uplift and a river incising stone told dramatically in the Canyon of Lodore and throughout the Grand Canyon; the story of a floodplain that gradually rises above the channel. The river tells many stories as it traverses the variable landscapes between its source and its destination.

Nowhere in the United States are these many stories of the interaction between rivers and regional geology more clear than in the Colorado River. Here the thin skin of sparse vegetation cannot hide the underlying rocks. Here the earth's anatomy is visible over an unusually wide expanse of land. We see, not only the venous network of the river and the behavior of the fluid it carries, but its effect upon the structures that the river courses around and over and through.

*

The Colorado River's headwaters are the Rocky Mountains of

Colorado and Wyoming, and the high plateaus of central Utah. Every large tributary of the Colorado River flows from high country across lower-elevation deserts: the Grand River (which is now called the upper Colorado River), the Green River, the Colorado River (which forms at the confluence of the Grand and the Green rivers), and the San Juan River.

By the time the water crosses the Arizona border and enters the Grand Canyon these large rivers have merged their identities into the one mighty flow known as the Colorado River. Downstream from the Grand Canyon, the Colorado continues its southwesterly journey, shaping the borders of southern Nevada and California, flowing through increasingly arid basins and ranges until it reaches the Sea of Cortez.

The modern Colorado River Basin crosses three very different landscapes—the Rocky Mountains, the Colorado Plateau, and the Basin and Range. These landscapes have diverse geologic histories that encompass the creation and composition of the rocks as well as their subsequent burial, deformation and uplift—enormous changes occurring over vast spans of time ultimately resulting in the linkage of the various segments of the Colorado River system into one drainage basin.

The basin began to take shape about sixty-five million years ago when rocks were deforming into great up- and downwarps of the earth's crust. Some segments of the Colorado River, which came into existence then, flow in nearly the same locations now, tens of millions of years later. But these very old segments coursed through a landscape that only faintly resembles the modern western United States. Some drained to the Pacific or Atlantic Oceans or into large lakes, and they were not linked together.

In the last ten or twenty million years, the entire Colorado Plateau and Rocky Mountain regions have been uplifted about one mile above sea level. This uplift has initiated a process that geologists refer to as

Facing page, top: Meanders on Colorado River as it heads into the Gore Range at Kremmling, Colorado.
Hay farmers take advantage of the Colorado's floodplain just before the river dives into Gore Canyon.

Facing page, bottom: Flat Tops near Derby Peak, Colorado.
The Flat Tops are a broad high country formed by basalts that covered the mountains above Glenwood Springs, Colorado.

Above: Farms on the Colorado River above Burns, Colorado.

Facing page: Colorado River at Sweetwater, Colorado.
Farms cling to terraces and are squeezed into narrow valleys along the river. The surrounding country is too steep and too dry to farm, and is used instead for grazing.

Above: Derby Creek above Burns, Colorado.

rejuvenation, during which thc basin's rivers have incised their courses into some of the very old up- and downwarps. Rejuvenation also caused the disparate river segments to link together into the Colorado River system that we know today.

Beginning about ten million years ago, North America had moved so far to the west that the continent began to override an area of high heat flow and upwelling from the earth's mantle. That heat caused thinning and stretching of the crust, and much of present-day Nevada, southeastern California, and southern Arizona broke into a series of north-south trending mountain ranges and basins. Farther south, the Sea of Cortez opened into an arm of the Pacific Ocean, giving the river its modern outlet. Since then, vast quantities of eroded sediment have been delivered to the lower Colorado River valley, to its delta, and finally to the sea.

*

The large rivers of the Colorado River Basin can be generally distinguished into three distinctive types: rivers meandering through wide valleys, rivers meandering through narrow canyons, and rivers whose course is partially blocked by debris flows.

Where a valley is wide and the channel slope flat, the river meanders in wide loops. The banks are typically sandy, overgrown with dense vegetation such as saltcedar, also known as tamarisk, and willows. Large sandbars sometimes occur along alternating banks of the river, and these bars are crossed by shallow, ever-changing chute channels. At flood stage, the bars are entirely covered by the river and are scoured and filled by the massive volume of rushing water and sand. At low-river stages, the secondary channels often become areas of stagnant flow. Within the widest valleys, closed-depression lakes may be found. Large, permanent islands typically fill parts of the channel.

If rapid geologic uplift occurs beneath a meandering river, the twists of the river may be locked into stone as the walls of rock rise and the river cuts down. Dramatic incised meanders are found within Canyonlands National Park and on the San Juan River below Mexican Hat, Utah. In these very narrow canyons, there is no room for floodplain lakes, and islands are few. Here the floodplains build by a stacking of sediments in a process referred to as vertical aggradation. Layer by layer, these floodplains build ever higher above the stream bed until they can only be overtopped by rare, very large floods.

The third type of river also flows through a very narrow canyon, but is susceptible to an event that distinguishes it from other sections of the river. During periods of unusually wet weather, normally dry sidestream channels may rage with flash floods and mixtures of water, sand, and rock that resemble flowing concrete. Debris flows constrict the main river channel, and their large accumulations of boulders—some as large as houses—create rapids with a distinctive pattern of streamflow upstream and down. Here debris fans shape the channel into a succession of rapids and eddies. Debris flows are common in Grand, Marble, Desolation/Gray, and Cataract Canyons, the Canyon of Lodore, as well as other canyons in the Uinta Mountains.

*

Geologists describe a region not only by the age of its rocks but also by the time when the mountains and plateaus formed. These periods of uplift necessarily occurred much more recently than creation of the rocks themselves. The borders of the Colorado River are marked by great uplifts of varying age—the Wind River Range in Wyoming, the Rockies and the volcanic San Juan Mountains in Colorado, and the Wasatch Plateau in Utah. Within the basin, there are lesser but still impressive uplifts such as the Uinta Mountains, the Kaibab Plateau, and the San Rafael Swell. These warps of the crust occurred during the Laramide Orogeny as western North America collided with the sea floor of the Pacific, compressing the continent's western interior. Some of the greatest spectacles of the basin are found where rivers have cut deeply into these uplifts, such as at Grand Canyon and the Canyon of Lodore, and at Split Mountain below Lodore, where the Green River sliced right through the center of a broad dome of rock.

Facing page: Black Canyon of the Gunnison National Monument, Colorado.

Above: Colorado River above DeBeque, Colorado.

Right: Colorado River near Rulison, Colorado.
Above their confluence in Utah, the Colorado and Green are two very different rivers. The Green generally flows through high, dry country that has always been difficult to farm and is sparsely populated. The Colorado's basin, on the other hand, has been more easily settled. Interstate 70 and the Denver and Rio Grande Railroad bracket the river's course from Burns to the Utah border.

Above: Grand Valley at Fruita and Grand Junction, Colorado.

The character of the river and its surrounding scenery is partly determined by the character of the rocks in which the river has eroded its course. The oldest of these rocks, such as the granites, schists, and gneisses exposed in parts of Grand Canyon and Westwater Canyon, are nearly half the age of the earth. Thick sequences of sedimentary rocks exposed elsewhere, such as in the Canyon of Lodore and in the Big Bend area of Grand Canyon, are somewhat younger, but still pre-date the occurrence of abundant fossil life. Other thick sequences of sedimentary rocks, formed during the evolution of abundant fossil life, are now exposed throughout the Colorado Plateau.

A river cutting through a major geologic structure encounters many types of rocks. Hard rocks, being more resistant to erosion, will produce a narrow canyon; weak rocks, inversely, will result in a wide valley. But when a river encounters hard and then weak rocks as it cuts through preexisting uplifts and downwarps, it will develop a sequence of narrow canyons and wide valleys that could not be predicted by an understanding of river processes alone. The river has shaped its own course, but along a path that has been guided by the rocks themselves.

Canyons are narrow and deep where rivers cross the hardest rocks, such as the very old, dark gray-and-black rocks of Upper Granite Gorge in Grand Canyon and Westwater Canyon in Utah. Hard, red sedimentary rocks, such as those in the Uinta Mountain Group that are about 800 million years old, form the narrow Canyon of Lodore, but similar aged, softer, mudstones in the Big Bend area of Grand Canyon have eroded

to form a relatively wide part of this canyon. Resistant layers of limestone, interbedded with some thick sandstones, form many of the cliffs in Grand Canyon and Cataract Canyon. Conversely, weak, gray marine shales erode to form gently sloping badlands near Green River, Utah, and Grand Junction, Colorado. In these areas, the valley is very wide and the river meanders with a gentle gradient.

*

The Green River, the longest tributary of the Colorado, provides a case history of the interplay between geology and river that probably occurred on the other tributaries of the Colorado, such as the upper Colorado and the San Juan. The Green, however, was more accessible, and consequently attracted the earliest interest of trappers as well as geologists. John Wesley Powell, who began his expeditions on the Green in 1869, was able to ship his boats to the Union Pacific's bridge crossing at the town of Green River, Wyoming.

The Green River begins its course in Wyoming's great granite range, the Wind River Mountains on the slopes of Gannett Peak, flowing as a wonderful meandering brook through dense forest, gravel bars, and deep pools. The valley is narrow and very deep, and bears the gouges of glaciers that covered the Wind Rivers as recently as about 15,000 years ago.

In the Bridger Wilderness, the backpacker walks the gentle valley of the Green but faces a steep thousand-foot climb when leaving the glacial trough to view the craggy peaks of the Continental Divide. Farther downstream, the canoeist looking up from Green River Lakes at the edge of the range is treated to one of the greatest scenes in the American West: Square Top overlooking the lakes. Downstream from the lakes, the Green River has a wide and shallow gravel bed, home to trout and haven to fly fishermen. The Green and its tributaries—New Fork, Big Sandy, Hams Fork, and Blacks Fork—are among the West's most productive gravel-bed trout streams. These streams join one by one, flowing down the edges of mountains into the center of the Green River basin. The river is where geologists would expect it to be—in the middle of a structural depression of downwarped rocks.

The river's bed becomes more sandy downstream, the bars become larger, and the river evolves into a sedate lowland stream, just as we would expect of an orderly river. Here wagon trains on the Oregon and Mormon Trails forded the Green and here, too, the transcontinental railroad bridged the river in 1868. A few early trappers and California-bound travelers tried to float on down the Green, thinking it an easier alternative to crossing the rest of the continent by land. But these explorers in bull boats and flat rafts soon discovered what Powell would later confirm in 1869. The Green River quickly turns into a maelstrom as it encounters the northern upturned edge of the Uinta Mountains. The river leaves its open basin and enters an ever deepening canyon at Flaming Gorge, emerges briefly in Browns Park, then turns sharply to dive straight into the towering walls of the Uintas at the Gates of Lodore.

Geologists believe the Green River once flowed east, parallel to the Uinta Mountains, to eventually join the Missouri River. Subsequent burial of the Uinta Mountains beneath their own erosional debris allowed the river to establish a southerly course. But as the mountains were uplifted

Facing page: Colorado River at the head of Westwater Canyon, Utah.

Left: Colorado River at Westwater Canyon, Colorado. The Colorado has cut down to Precambrian schist that is arched up beneath Westwater Canyon. The overlying tan cliffs are Mesozoic shales and sandstones that erode away more easily than the darker schist.

Below: Colorado River below Westwater Canyon, Colorado.

48

Above: Red Mountain in the San Juan Mountains south of Ouray, Colorado. The headwaters of the Uncompahgre River gather in the high San Juans above Ouray, Colorado, where mining flourished in the 1870s. The river now runs orange with a mine's tailings.

again, the river was trapped and began to erode into the eastern shoulder of the Uintas.

After the modern river leaves the Uinta canyons, it crosses the broad structural Uinta Basin composed of weak, easily eroded rocks. The river meanders in a wide valley, its cross-section wider here than anywhere else along its course. Farther south, the Green flows through a backward-tilted layer cake of sedimentary rocks as it flows through layer after rising layer of rock—the hard strata spawning canyons such as Desolation, Gray, Labyrinth, and Stillwater, and weaker strata creating wide valleys, such as that at the town of Green River, Utah.

At last the Green joins the Colorado just above Spanish Bottom in the heart of Canyonlands National Park in Utah, flowing now southwestward through alternating weak and hard rocks in Cataract, Glen, and Marble canyons. Below its confluence with the Little Colorado River, the Colorado turns west and flows through the Inner Gorge schists and granites of the Kaibab Plateau in Grand Canyon.

Above: Gully erosion in the Paradox Valley at Bedrock, Colorado.

Drainages continually adjust themselves to cycles of erosion and deposition. If climate and river flows suddenly change or if a valley loses its vegetative cover, erosion will carve steep gullies. A cycle of such erosion swept through the Colorado River Basin in the 1880s about the time that grazing was being introduced.

Right: Cliffs above the Dolores River at Gateway, Colorado.

Above: Cottonwood Wash and the Grand Valley north of Cisco, Utah.

At the Grand Wash Cliffs, the river reemerges from its deep, labyrinthine journey through Grand Canyon. It was here, on his first night beyond its awesome and intimidating presence, that Powell was moved to write, "danger and toil were endured in those gloomy depths, where ofttimes clouds hid the sky by day and but a narrow zone of stars could be seen at night. Now the danger is over, now the toil has ceased, now the gloom has disappeared, now the firmament is bounded only by the horizon and what a vast expanse of constellations can be seen!"

*

In the decades since Powell made these rapturous observations, the Colorado and its tributaries have been largely subjugated to the service of human needs and wants. Diversions of water from the upper Colorado River Basin and large dams on the main channels have significantly transformed the hydrology of streamflow and the transport of sediments through the basin system.

The stream channels themselves have changed, too. The magnitude and style of such change—either erosion or aggradation—are dependent upon the location and size of each dam, the amount of sediment trapped behind it, and the characteristics of the downstream channel.

In canyons affected by debris fans where sediment is accumulating, eddies fill, and the distinctive topography of eddy sandbars is lost as vegetation invades these areas. This type of channel-filling is evident in the Canyon of Lodore and to some degree in Desolation Canyon. In contrast, in debris-fan affected canyons where sediment is being eroded away, large eddy sandbars may be completely removed and replaced by open

Facing page: Colorado River through Professor Valley upstream from Moab, Utah.

Above: White Ranch on the Colorado beneath the La Sal Mountains and Castle Valley near Moab, Utah.
An irrigated field now occupies an alluvial fan at the foot of Castle Valley. The fan has forced the river toward the bottom of the picture, momentarily steepening its gradient and creating a rapid.

expanses of water. Campsites prized by recreational boaters gradually disappear. In meandering reaches, aggradation causes channel narrowing, elimination of islands, and loss of topographic diversity on floodplains. Where meandering reaches are depleted of sediment, banks erode and channels incise.

What does the future hold? The answer depends to a great extent upon us—upon how we operate our dams and whether we choose to remove a few; upon how much water is withdrawn from the basin for growing crops and cities. Other factors affecting the basin are beyond our control—how future climate change might affect natural processes of runoff and sediment delivery, for example. Whatever their cause, results of these changes will undoubtedly be profound and wide-ranging.

NED ANDREWS WRITES ABOUT THE ORIGIN OF WATER AND SEDIMENT WITHIN THE BASIN, HOW FLOWS HAVE BEEN MODIFIED, AND THE EFFECTIVENESS OF RESERVOIRS IN SUPPLYING REGULATED WATER TO USERS IN AND OUT OF THE BASIN.

Wet River, Dry River

E. D. Andrews

Facing page: Jointing in the Navajo Sandstone above Salt Wash, Arches National Park, Utah. The Navajo Sandstone has been bent upwards north of the Colorado River at Moab, Utah. In the process, the sandstone was systematically fractured and now weathers into fins and erodes into arches.

Above: Salt River flood.

THROUGH CYCLES OF DROUGHT AND FLOOD over millions of years, the Colorado River has carried the silt, sand, and debris of the American Southwest from the Rocky Mountains over fourteen hundred miles out to the Sea of Cortez. The magnitude of its impact on this landscape is clearly evident from the awesome names the river has inspired: Grand Canyon, Gates of Lodore, Split Mountain, Monument Valley.

During the last five decades—a geologic millisecond—the rock-crushing power of the Colorado ground to a halt behind concrete walls. Dams placed on the river at Black Canyon, Glen Canyon, San Juan Canyon, and Flaming Gorge now store and divert water for cities, farms, and ranches. The entire natural flow of the river is consumed above its lower reaches, and the winter snows of western Colorado no longer make their way to the Pacific Ocean. Now for the first time since the Rocky Mountains rose out of the North American plains, no water runs out of the Southwest to the Sea of Cortez.

Ironically, as half a century of dam building comes to a close, the hydrology, geomorphology, and ecology of the Colorado River are becoming increasingly important in the political landscape. The Colorado River drains the third largest river basin in the continental United States, following only the Mississippi and Columbia in size. It and its tributaries flow through eighteen national parks and monuments. Whitewater boating has become a major commercial activity, with more than twenty thousand people floating through just the Grand Canyon each year. Increasingly intense competition has developed for agricultural, industrial, and municipal water to supply growing Southwestern cities. Meanwhile, five of the Colorado's native fish are listed as threatened or in danger of extinction, and water is needed for instream flows to maintain the river's historic flora and fauna. There is no unused water in the Colorado Basin to meet these future demands—rather, water must be taken from one

Facing page: Evaporation ponds at Potash, Utah, downstream from Moab.
The Paradox Formation is a thick salt-rich stratum beneath the Colorado River in eastern Utah. Salts are recovered by pumping water into the stratum and recovering the resulting potassium-rich brines. Evaporation then yields concentrated minerals ready for market.

Above: Atlas Mill uranium processing site on the Colorado River at Moab, Utah.
The Colorado River laps up against the old Atlas Mill site near Moab, Utah, where uranium was processed in the 1950s and 1960s. The tailings have been stabilized but still leach significant quantities of heavy metals into groundwater and the river.

user to serve another. To balance these demands, the river's managers must develop a more sophisticated understanding of the basin's streamflow variability and hydrologic resources, along with its extensive system of reservoirs and diversions.

*

The Colorado River Basin is arid; most of it receives less than twenty inches of precipitation per year and much of it, less than ten inches, qualifying it as one of the earth's most dramatic deserts.

A comparison with the Mississippi River helps put this aridity into perspective. At 244,000 square miles, the Colorado River Basin is about one-fifth the size of the Mississippi, but the Colorado River carries only one-thirtieth as much water.

Not only are the Colorado's flows relatively small for its drainage area, they vary greatly from season to season and year to year. Most of the river's flows come from headwaters near the crest of the Rocky Mountains. These waters fall as snow in the winter, and are then released to the basin in springtime, transforming the snowpack to a raging torrent. The natural flows of the river from May through August are twenty to thirty times the flows expected for September to April.

How much annual runoff has flowed past Lees Ferry, at the entrance to Grand Canyon, over the past several hundred years? Unfortunately, the answer is not directly available, since our records of the river's flows date only to 1895, and one hundred years of flow records offer no assurance that we have documented the river's extreme floods and droughts. Have the last one hundred years, for example, shown us a drought of the

Left: San Rafael River flowing through the San Rafael Swell west of Green River, Utah.
The San Rafael Swell is the high country above a monocline, where rocks were smoothly rolled into an arch during the Laramide Orogeny, 65 million years ago. The San Rafael River and its countless small tributaries now dissect the Swell, carrying it away grain by grain.

Below: San Rafael River at Dugout Wash south of Green River, Utah.
The San Rafael is a permanent stream in the middle of a terribly dry and otherwise forbidding desert.

Above: Green River above its confluence with the San Rafael River, Utah.

magnitude that drove the Anasazi from the Four Corners region in the twelfth century? Yet hydrologists have ways to extend the period for which flow information is available and with some scientific sleuthing, they can construct a good estimate of the river's historic water power.

The most successful approach to understanding ancestral flows in the Colorado River Basin has come from the study of annual tree-ring growth. Using a bore, scientists extract a one-quarter-inch-wide core of wood from a tree that is several hundred years old. The tree's annual growth rings are dated, and their width measured. A mathematical relation between the tree-ring width and annual runoff is determined from recent data. From this relation, knowing the width of a growth ring that was added in a given year allows an estimate to be made of that year's water runoff.

Using tree-ring studies, annual runoff records have been reconstructed back into the past well beyond 1895. The most remarkable feature of this reconstructed record is the sustained periods—often twenty to fifty years—when the river consistently flowed above or below average. This phenomenon is called persistence, and it is common to most streamflow records worldwide.

Persistence refers to global weather patterns that, once established, tend to recur. Dry years follow dry years, and a sustained period of below-average runoff is a drought. Following an especially dry year, the amount of water naturally stored in a watershed is relatively small. Conversely, after an especially wet year, soil moisture is high, aquifers and lakes are full, and snow fields may be large enough to last until the next

winter. Furthermore, each year's successive storms generally follow a similar path, leading to significant snow accumulations in wet years.

*

Developing a dependable water supply in a region characterized by extreme climatic variability requires governments to hold the Colorado's flow behind dams during periods of high runoff, and then release it during periods of low flow. Beginning in the early 1930s, a dozen large dams and reservoirs were constructed along the mainstem of the Colorado and all its major tributaries. Reservoir capacity in the basin now exceeds sixty-three million acre-feet. If the water held in full reservoirs were gradually released, it would be enough to replace four years' worth of the river's average annual runoff.

The storage and regulation of river flows have substantially increased the dependable water supply for human use. Flow regulation and depletion, however, have dramatically altered the river and its tributaries.

*

Nowhere along the Colorado River is the conflict between the development of water supplies and riverine resources more obvious than through the Grand Canyon. Glen Canyon Dam, which forms Lake Powell, is located fourteen miles upstream from Lees Ferry and the boundary of Grand Canyon National Park. Because it captures nearly all of the Rocky Mountain snowpack, Glen Canyon Dam regulates the flow of water to the lower Colorado River Basin.

Before Glen Canyon Dam blocked the Colorado River in 1963, the Grand Canyon's peak flows occurred during floods in May and June. The average pre-dam flood approached one-hundred thousand cubic feet per second. After the floods subsided, however, the flow in the Colorado typically fell to as little as three thousand cubic feet per second, except during brief, but occasionally substantial, tributary flash floods. The largest flash floods raised the water in the main channel to the magnitude of the snowmelt peaks.

The storage of water in Glen Canyon Dam has greatly reduced the annual range of river flows. Since 1962, the average "flood" discharge has been only 33,800 cubic feet per second—less than one half the historic mean. Stored spring runoff is released during the remainder of the year, so that the volume of water released in any given month now varies only by a factor of about two. This has substantially increased river levels during the previously low-flow months of September through April.

The primary purpose of Glen Canyon, and other dams in the Colorado River Basin, is to regulate river flows to fulfill beneficial uses and to apportion water to consumers throughout the basin. Generation of hydroelectric power is an important secondary purpose of the Glen Canyon Dam's operations. Water released from the dam varies hour to hour and day to day in response to the demand for power throughout the western United States. Typically, this demand is reduced on weekends and at night. Consequently, weekend releases are much less than weekdays, and daily flows are higher than at night.

The elimination of natural spring floods and the daily regulation of water flows have had dramatic effects on the Colorado River channel and the native aquatic and riparian ecosystems. The size, sinuosity, and

Facing page, top: Green River at Fort Bottom, Canyonlands National Park, Utah.
The Green River flows south into the rising country of Canyonlands National Park.

Facing page, bottom: Millard Canyon joining the Green River at Queen Anne Bottom, Canyonlands National Park, Utah.
Inside Canyonlands National Park, the Green River is locked between walls of resistant sandstone. The river adjusts its course within these walls, alternately depositing and eroding sandbars depending on flood frequency and intensity.

Facing page, top: Confluence of the Green and Colorado rivers, Canyonlands National Park, Utah.
Two great rivers join inside Canyonlands National Park. The Green is actually the longer of the two. The Upper Colorado was known as the Grand River until just before the 1922 Colorado River Compact. The Colorado state legislature had changed the Grand's name to give better leverage during negotiations that would determine which state received how much of the system's total flow.

Facing page, bottom: Colorado River at the Loop, Canyonlands National Park, Utah.
This meandering pattern was inherited from an earlier, lazier geomorphology, incising as the river cut down and became trapped between bedrock walls. Eventually the river must cut through the necks of these loops to shorten its course.

Above: Colorado River at Little Bridge Canyon, Canyonlands National Park, Utah.

shape of a river channel is generally determined by a few days of high flow each year. When floods are substantially reduced, the channel will adjust over a period of years to the new conditions. Sediment fills the channel. Riparian vegetation encroaches upon the river banks. The removal of sediment changes the clarity and color of the water. Native fish which were adapted to survive in large muddy floods lose their adaptive advantage over non-native fish which, before the dam, had been swept away by the torrents. All of these changes have been observed in the Grand Canyon since the gates closed on Glen Canyon Dam.

During the spring of 1996, an experimental flood was released from Glen Canyon Dam to attempt to reverse some of these effects. For one week, forty-five thousand cubic feet per second of water was released through Grand Canyon—not nearly the magnitude of a high-year flood before the dam was built, but significantly more than the regulated flows to which the river had become accustomed. The experimental flood was intended to restore critical river channel features along with native aquatic and riparian habitat. Preliminary results show that the reintroduction of a substantial flood to the river's life temporarily reinvigorated its beaches and provided greater opportunities for native willow to compete with introduced tamarisk. But within a year of the experiment, more than eighty percent of the new beaches were gone. Because of the feared impact of recurrent floods upon downstream users and power generation,

Muddy Creek inside the San Rafael Reef southwest of Green River, Utah.

Muddy Creek as it enters the Sinbad country northwest of Hanksville, Utah.

it remains to be seen whether traditional flood flows will be added to the on-going repertoire of management of Glen Canyon Dam.

*

The system of reservoirs in the Colorado River Basin was designed, constructed, and is now operated by the U.S. Bureau of Reclamation. Such an extensive program would not have been possible without the consent and agreement of the seven states of the Colorado River Basin. The states' interests differ, though, related in part to the speed at which each has developed uses for its share of available water.

The bureau's control of the river basin arose from a political squabble in the early part of this century. In 1901, farmers in the Imperial Valley of California began diverting substantial quantities of water from the Colorado River, and it appeared that soon they would be able to call for, and use, all the water that could be channeled into their ditches. Upstream states became concerned that California's continued agricultural expansion would preempt their opportunity to eventually use a portion of the basin's water. Consequently, the seven states entered into a compact in 1922 to divide the river's flow between the Upper Basin (Wyoming,

Above: Rainshower above Muddy Creek at Cedar Mountain, Utah.
Much of the moisture that falls within the Utah desert arrives during brief, intense thunderstorms. The resulting heavy runoff has much more erosive power than that provided by a slower, gentler rain.

Right: Aspens at the headwaters of Pine Creek on the Aquarius Plateau, Utah.

*Above: Waterpocket Fold in Capitol Reef
National Park, Utah.*
*The Navajo Sandstone (capped by the Carmel
Formation) has been lifted with the
Waterpocket Fold and subsequently dissected by
innumerable streams within Capitol Reef
National Park.*

Colorado, New Mexico, and Utah) and the Lower Basin (Arizona, California, and Nevada).

The Colorado River Compact by no means resolved all of the issues that divided the states. But it did function well enough to allow the construction of Hoover Dam, named for Herbert Hoover, who led the successful compact negotiations.

In 1922, as serious negotiations between the states were underway, it appeared that the average flow of the Colorado River at Lees Ferry (within a few miles of where the river crosses from the Upper to the Lower Basin) was slightly more than sixteen million acre-feet per year. Surviving records of the negotiations suggest that the states intended to divide an annual flow of fifteen million acre-feet evenly between the Upper and Lower basins, while reserving approximately one million acre-feet a year to meet an anticipated future request from Mexico. Instead of dividing this flow equally, however, the Upper Basin states agreed to release an average of 7.5 million acre-feet per year to the Lower Basin, plus half of any future allotment for Mexico. This agreement placed all of the rewards—and risks—of hydrologic variability on the Upper Basin states. The Lower Basin states chose certainty (a ten-year average of 7.5 million acre-feet a year), while the Upper Basin states chose to gamble on whatever remained after they released 7.5 million acre-feet.

As it's turned out, the Lower Basin states chose the better course. The runoff expectations which drove the compact negotiations were not based upon typical years of flow. The period from 1899 to 1921, which was the only information available when the compact was negotiated,

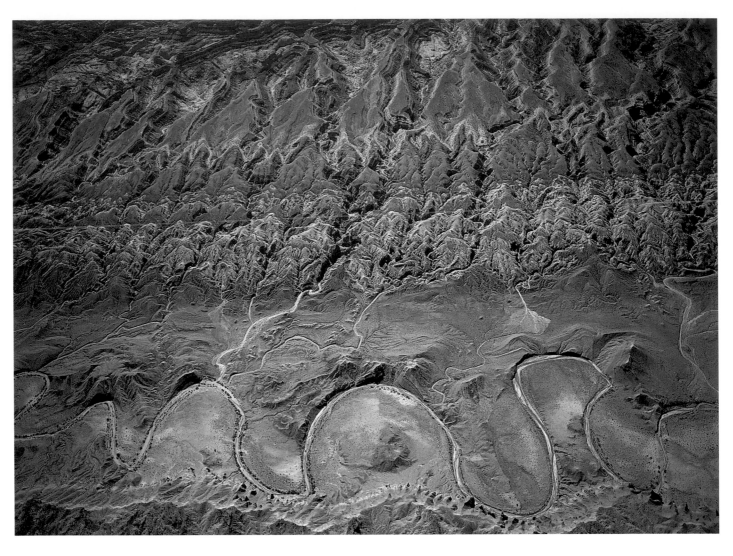

Facing page, top: Deep Creek, Capitol Reef National Park, Utah.
North of the Fremont River, the Waterpocket Fold is carved by gullies that rush down to join the meanderings of Deep Creek.

Facing page, bottom: Mancos Shale south of Caineville, Utah.
Life has a hard time establishing itself amidst the inhospitable Mancos Shale. The resulting moonscape, lacking vegetative protection, erodes quickly.

Above: Mancos Shale along the North Caineville Reef, Utah.
The North Caineville Reef is yet another monocline that tilts sedimentary rocks east of Capitol Reef National Park. Resistant sandstones are left standing above the surrounding shales as erosion preferentially removes the softer strata.

included one of the highest persistent runoff periods in the past 450 years. With records reconstructed from tree-ring analysis, we now know that the average annual runoff for the entire period from 1530 to 1990 is 13.7 million acre-feet, far less than the sixteen million acre-feet expected when the compact was negotiated.

✳

It is now clear that the Colorado River Compact presumed that more water would be available over time than the river can actually deliver. Moreover, the current storage capacity of the Colorado River Basin has already surpassed what is optimal for the regulation and delivery of water for consumptive and instream uses. Indeed, the addition of new reservoirs designed to support population growth in the Southwest could actually decrease the net volume of water which can be used in the basin, because the additional reservoir water surface area would serve only to increase water losses through evaporation and bank seepage. It is hard to escape the conclusion that as the face of the Southwest changes, the Colorado River's managers and users will have to make hard choices about the reallocation of existing water supplies, rather than invoking increased storage capacity in response to growing—and competing—water demands.

Water, Earth, and Sky: The Colorado River Basin

RICH VALDEZ DESCRIBES THE LIFE AND
TIMES OF NATIVE FISH WITHIN THE
COLORADO RIVER BASIN.

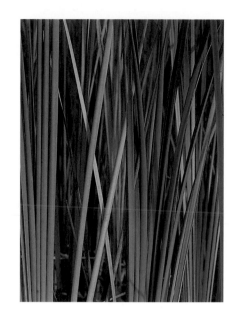

Of Suckers, Chubs, and 100-Pound Minnows

Richard Valdez

*Facing page: San Juan River at Chinle Wash
downstream from Bluff, Utah.
Chinle Wash joins the San Juan River as it
flows along the flanks of the Comb Ridge
monocline. Chinle Wash adds a heavy
concentration of fine sediment derived from the
Navajo Reservation within Arizona and
southeastern Utah.*

Above: Reeds at Taylor Lake, California.

THE SOUTHWEST UNITED STATES hardly seems a hospitable place for fish. The land is arid and harsh, and what little water exists is often muddy and salty. Yet, the Southwest's major river, the Colorado, supports a native fish fauna with an evolutionary history dating back three to five million years. A limited number of fish species evolved in the Colorado River because a long period of geographic isolation minimized genetic exchange with other river basins. Also, a rigorous environment—marked by large seasonal variations in water flow, temperature, salinity, and sediment load—created conditions favorable only to few fish. Those that did survive are often described as ugly or bizarre, but all are remarkably adapted, with slender, hydrodynamic bodies, dorsal humps, and large fanlike fins for swimming in swift currents; thick leathery skin and few scales to resist the abrasive sediment; and a highly developed lateral line and sensory system to navigate and locate food in the dark waters.

Equally remarkable as the individual species is the assemblage of fish within the Colorado River Basin. Seventy-four percent of its thirty-five species are indigenous and found nowhere else on earth. Although eleven families of fishes are represented, twenty-three of the thirty-five species belong to one of two families—Cyprinidae (minnows) and Catostomidae (suckers). Some are particularly impressive, such as the 100-pound Colorado squawfish, the world's largest minnow.

The native fishes of the Colorado River fit into four major categories: big-river fishes, mostly endemic, that once ranged throughout the mainstem and larger tributaries; small to medium-sized forms, largely endemic, that occupied smaller tributary streams at low to intermediate elevation; cool and cold water forms that occupied high to intermediate elevations with close relatives in adjacent basins; and estuarine and marine forms that once occupied the delta and lower river.

*

Facing page, top: Comb Wash and Comb Ridge northwest of Bluff, Utah.
Comb Wash drains the eastern edge of Cedar Mesa, home to dense settlements of the Anasazi until 700 years ago.

Facing page, bottom: Lake Powell, looking from Hite toward the Henry Mountains.

Above: Lake Powell and the west side of Cummings Mesa, Utah.

The Colorado River has seen dramatic geomorphic changes in the last five million years, including the carving of the Grand Canyon, a major redirection of flow in the lower basin, and numerous large lava flows that dammed the river for many years. Nevertheless, many of the fishes living in the river today originated during this period of change. And though some forms failed to survive, the fish fauna changed little until the advent of human interference. In the large, muddy, and often turbulent waters of the mainstem were found razorback suckers, flannelmouth suckers, bonytails, roundtail chubs, humpback chubs, and the only major predator, the Colorado squawfish—all endemic, big-river fishes. In the smaller tributaries, there were woundfin, Virgin spinedace, spikedace, loach minnows, speckled dace, bluehead suckers, and desert suckers. Higher-elevation streams supported Colorado cutthroat trout, Gila trout, mountain whitefish, mountain suckers, speckled dace, and mottled sculpin. In the lower river, spawning runs of searun striped mullet and

Above: Navajo Canyon area, Aztec Creek, Arizona.

Right: Flashflood, Courthouse Wash at its confluence with the San Juan River, Mexican Hat, Utah.

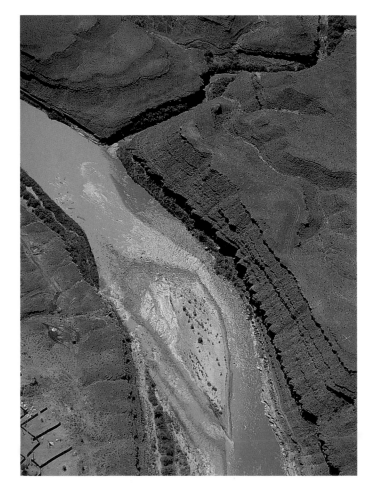

Pacific tenpounders made their way up from the Sea of Cortez, and the marine toatuava deposited its eggs and reared young in the rich estuaries of the delta. The diminutive desert pupfish survived in seasonally flooded riverside ponds and springs of the lower river. Despite its cataclysmic history and harsh appearance, the Colorado River Basin supported a thriving collection of fish.

The Colorado River was highly variable, sometimes overflowing its banks in springtime and spreading far across the land. In other years, great droughts reduced the river to a salty trickle. This would seem to be a harsh, unpredictable environment for fish. Yet it was the unpredictable nature of the river that shaped many of its fishes. The survival and well-being of each fish species in the Colorado River Basin was highly dependent upon the harmonious occurrence of many environmental factors, each necessary for a particular life stage. The river's spring floods re-shaped nursery habitats, delivered annual doses of food-producing nutrients, and scoured spawning areas of muds and silts that otherwise smothered eggs and young. Late summer monsoon floods revitalized production and helped transport growing fish downstream to warmer, more productive, food-rich areas. One silt-laden flood at the wrong time, however, could suffocate the eggs of an entire class, postponing the prospect of a new generation to yet another year. Evolution prepared these fish for such failures with unusually long life spans—Colorado squawfish, for example, live over forty years. One successful spawning year in ten was all that was necessary, perhaps, for enough young to survive to perpetuate a species.

During spring floods on the Colorado River, cutthroat trout spawned at high elevations, while large congregations of flannelmouth suckers and bluehead suckers gathered to spawn in tributaries and inflows. Razorback suckers massed over large mainstem gravel bars, each female depositing fifty to one-hundred thousand tiny, sticky, pinhead-sized eggs. The young razorbacks hatched after incubating only five days as tiny slivers of finless translucent flesh, visible only as two sand-grain-sized eyes. Only about one-third inch long at hatching, the emerging larvae were swept downstream into flooded riverside bottomlands. Though temporary, these shallow protected environments were rich with tiny crustaceans— rotifers, copepods, and cladocerans—each succeeding the other and each increasingly larger in size to satisfy the needs of the hungry, growing fish.

Suspended in the water column to avoid stranding in mud puddles, young razorbacks were carried back to the main channel on receding flows. When the fish reached slightly over one inch in size, their mouths migrated from the end of the head to a permanent position on the underside of the head. Their feeding switched from midwater crustaceans, not plentiful in main-channel habitats, to bottom-dwelling insects, algae, and nutrient-rich river sediments. By midsummer, the young razorbacks were two to three inches long and lived in shallow riffles and near-shore backwaters and embayments. Adulthood came after two or three years, and the fish migrated long distances before spring runoff to reach common spawning grounds.

In mid- to late summer, as the spring floods receded, large Colorado squawfish—some five to six feet long and weighing eighty to one-

hundred pounds—migrated hundreds of miles upstream to reach large mid-elevation spawning riffles, recently cleansed of sediments by spring runoff. Each female made multiple spawning forays in the company of several males, broadcasting twenty-five to seventy-five thousand tiny, sticky eggs, each less than one-tenth inch in diameter, into the security of well-oxygenated spaces between the cobble and gravel. Like those of razorback suckers, eggs of Colorado squawfish incubated for about five days, and river currents transported the emerging young downstream from their rocky shelters. Small and vulnerable, these tiny fish were swept into warm, protected, enriched nursery backwaters in sandy alluvial reaches. Squawfish matured in lower productive river reaches in three to five years before returning hundreds of miles upstream and resume a new cycle of life.

While Colorado squawfish and razorback suckers were highly migratory freshwater species, flannelmouth suckers, bluehead suckers, bonytails, and roundtail chubs were less mobile and able to satisfy their life-history requirements within smaller regions of the river. These fish also gathered in large numbers to spawn and broadcast their eggs over cobble and gravel riffles.

In marked contrast to these migratory species was the more sedentary and seemingly the most acutely adapted species of the assemblage, the humpback chub. Perhaps the most recently evolved of the big-river fishes, they did not tend to migrate, preferring specific canyon areas. Yet despite their propensity for whitewater regions, humpback chubs were poor swimmers; ironic for a species inhabiting the steepest, swiftest river canyons in North America. Perhaps other faster, better swimmers tried to evolve in the torrential Colorado River, but the evolutionary prize of survival went not to the swiftest, but to the fittest. Humpbacks reached a maximum length of only sixteen inches and weighed slightly over two pounds. The tapered body, large fanlike fins, enlarged dorsal hump, and flattened head enabled the humpback chub to "soar" with the ebb and flow of the river in low-velocity eddies—like an avian raptor soaring on thermal air currents—while feeding on entrained insects and vegetable matter delivered by the adjacent swifter river currents. The humpback found tiny struggling insects or lifeless bits of plant material in a river of mud not by means of enlarged eyes, but with a highly sensitized lateral line and a network of nerve endings, known as neuromasts, centered around the head, much like those of ocean-going sharks.

The humpback chub was also a spring spawner, aggregating in large numbers over mainstem gravel riffles or in tributaries shortly after the peak of spring runoff. Following a short incubation period of five days, newly hatched young made their way to nearby shorelines near their birthplace. Here they filled their air bladders, an adaptation allowing them to resist the currents that swept young razorback suckers and Colorado squawfish downstream. The sheltered shorelines provided food as well as protection from swift flows and predators; they remained among rocky slopes until maturity at about three years of age. Immature humpbacks had little or no dorsal hump, but the deposit of cartilage and fat quickly developed and grew through adulthood as a secondary sex characteristic. The hump could have been used as a hydrofoil to enhance stability in

Facing page, top: Monument Valley, Arizona/ Utah.

Facing page, bottom: Canyon of the Escalante River, Utah.
The Escalante River is at the heart of the newly created Grand Staircase-Escalante National Monument.

*Facing page: Lake Powell in Escalante Canyon
below Stevens Canyon, Utah.
These meanders first formed when the
Escalante River flowed across a gentler terrain.
Uplift shifted the balance of erosion and incised
the old patterns into bedrock. Ironically, the
balance has been shifted again, this time away
from erosion, by the creation of Lake Powell.*

*Above: Navajo Mountain and Lake Powell
southwest of Halls Crossing, Utah.*

the swift currents, or it may have served simply as a reminder to the fish
of its maturity and age.

*

Prehistoric Indians of the Southwest used the fish of the Colorado
River as food. Archaeological records from the lower Colorado River date
to as early as 300 B.C. to A.D. 400, and many date from A.D. 1100 to
1700, when increasing numbers of Native Americans lived in the region.
Remains from all the big-river fishes—Colorado squawfish, razorback
suckers, bonytails, and humpback chubs—have been found at
archaeologic sites. Partial skulls of Colorado squawfish from Catclaw Cave
indicate fish greater than five and one-half feet in length, weighing up to
eighty pounds. The fish were probably caught with woven dip nets or
shot with a bow and arrow, particularly in shallow flooded areas during
spring runoff.

Except as food, fish apparently have not played other roles in the
lives of contemporary Native Americans. Images of fish are notably rare

Above: Spires in the Claron Formation below Sunset Point, Bryce Canyon National Park, Utah.
Erosion has finely etched this terrain into the fragile pinnacles and spires of Bryce Canyon National Park.

Right: Braided channel of the Paria River near Cannonville, Utah.
The Paria River works this way and that across its floodplain, distributing sand here, eroding it there, striking a balance with the vegetation that would otherwise lock it into a single channel.

among petroglyphs on sandstone panels along many of the basin's waterways, panels that abound with images of other animals, such as bighorn sheep, bison, elk, and deer. While the Hopi Indians have a fish kachina, called Pakiowik, Colorado River fish appear to hold little ceremonial or religious significance to Native Americans in the Southwest.

The silty, torrential, and isolated Colorado River effectively hid its fish bounty from early explorers and even early scientists for many years. Records of early expeditions, such as those of Fathers Dominguez and Escalante in 1776, contain little about the fish. Colonel William H. Ashley gives us the first mention of the Colorado's native fish, when on May 27, 1825, he reported that his party had been subsisting on the Colorado squawfish caught in the upper Green River near the mouth of the White River. Ashley reported, "We find them of an excellent kind of a different speeces [*sic*] to any that I ever before have seen similar in appearance to our pike."

Many Colorado River Basin fishes were first described by North American ichthyologists S. F. Baird and C. Girard in the mid-1800s, from specimens obtained during regional military operations and the U.S.-Mexican Boundary Survey. Explorations of the region during the late 1800s and early 1900s revealed additional species and some information on distributions and abundance. Nevertheless, the humpback chub, a canyon dweller, was not described as a new species until 1946 from specimens in Grand Canyon, making it perhaps the most recent finding of a major fish species in North America.

*

Despite the cautions of Major John Wesley Powell, following his famed expeditions down the Colorado River in 1869 and 1872, land in the West quickly became partitioned under the Homestead Act, and extensive agriculture and livestock grazing reduced many stream courses to deep eroding banks and barren gullies. Natural habitats were further impacted starting in the late 1800s when the U.S. Bureau of Fisheries began to import exotic food fishes—primarily carp and channel catfish—to an otherwise "non-productive" West. The water, habitat, and neighbors of the native Colorado River fishes soon changed in dramatic and unprecedented fashion.

Noted ichthyologist R. R. Miller reported in 1961 that the last one hundred years had brought drastic changes to the rivers of western North America and their fish fauna as a result of human influence. Miller attributed the dramatic decline of native fishes to altered land-use practices, degraded water quality, and introduced species.

Prompted by concerns about flooding, small dams and diversions began to appear throughout the basin in the late 1800s. Beginning with Hoover Dam in 1935, fourteen mainstem dams were built on the Colorado River, converting much of its course into a series of quiescent pools. By 1963, large walls of concrete, some over seven hundred feet high, impounded the flow of the river and its tributaries: Flaming Gorge and Fontenelle Dams on the Green River; Navajo Dam on the San Juan River; Blue Mesa and Crystal Dams on the Gunnison; Roosevelt Dam on the Salt River; and the largest, Glen Canyon Dam, dividing the upper and lower basins of the Colorado. The Colorado River has been likened to a

Facing page: Cliffs and hoodoos of Entrada Sandstone east of Cannonville, Utah.

Above: Chinle Formation shales along Paria River northeast of Mollie's Nipple, Utah.

giant plumbing system, with its many reservoirs as holding tanks and the dams as spigots.

During this era of dam building, much of the river habitat was flooded or became fragmented, forever blocking passage for migrating Colorado squawfish and razorback suckers. The massive, salmon-like runs of the Colorado squawfish near Yuma, Arizona, where farmers would pitchfork thousands into horse-drawn wagons for use as fertilizer in the mid-1800s, came to an end. Also gone was the nearby cannery that distributed the Colorado River "white salmon" to markets across the country. By 1975, Colorado squawfish were gone entirely from the lower basin. The only reminders of this legacy are in memories of old-timers and in a few faded photographs. Grand Canyon river guides still recall a large log on the river bank used to mark the length of Colorado squawfish—some of which measured more than four feet. Attempts by biologists to re-introduce hatchery-reared Colorado squawfish into the lower basin have been largely unsuccessful—testimony that human-induced changes in

Facing page, top: Old Paria townsite on Paria River's cockscomb, Utah.

Facing page, bottom: Sand blowing into Paria River above the confluence with the Colorado River, Lees Ferry, Arizona.
Blowing sand is trapped in the Paria River Canyon above Lees Ferry. The sand is carried into the Colorado River at the Paria's confluence just below Glen Canyon Dam. The Colorado below Glen Canyon is a clear cold stream starved of any sediment by the dam. Much of the sand delivered by the Paria is deposited along the channel margins of the Colorado, awaiting floods that will distribute it into the bars that once graced the Grand Canyon.

Above: Paria River carrying sand to the clear Colorado River at Lees Ferry, Arizona.

only the last 150 years have exceeded those of five million years of geologic time.

To make room for the new array of game and food fishes in the reservoirs, state and federal agencies poisoned many miles of the Green and the San Juan Rivers in the early 1900s to get rid of "trash fish." Thus, squawfish, chubs, and suckers that were at the heart of the Colorado River ecosystem were now targets of a new management philosophy bent on replacing them with recreational sport fish. Taking the place of the native fish were largemouth bass, crappie, bluegill, channel catfish, and later walleye, striped bass, and smallmouth bass—all noted game fish and predators. Along with these new introductions came releases of forage species, such as gizzard shad, and extensive inadvertent releases of bait-bucket fish, such as red shiners, golden shiners, and fathead minnows. In 150 years, nearly seventy-five new species of fish were introduced into the Colorado Basin, many competing for food and space and bringing

Above: Triple Alcoves along the Colorado River in Marble Canyon, Arizona.

Right: Flashflood in Tsegi Canyon, Navajo National Monument, Arizona.
Sediment moves through the Colorado River system spasmodically. Brief intense floods that last for just a few hours will carry the greatest part of a year's sediment load. Deprived of these floods—either by drought or dams—a river's channel will clog with sediment.

new parasites and diseases that affected the already dwindling numbers of native fishes.

Downstream from the dams, the muddy river was replaced by clear, cold water released from the bottom of reservoirs. These conditions led to highly productive "tailgate trout fisheries" and excluded the native fishes from these regions forever. These blue-ribbon fisheries, famous world-wide for the numbers of trophy trout available to fly anglers, have become highly valued resources in the new river ecosystem.

The only wild Colorado squawfish left are in the upper basin's relatively free-flowing reaches of the Green, Yampa, and Colorado Rivers. Faced with an army of non-native predaceous and competing species, restricted habitat, and cooler, more restrictive growing seasons, Colorado squawfish persist only in low numbers. Hundred-pound specimens now seem a figment of the imagination. The largest Colorado squawfish caught since 1975 has been a twenty-five-pounder near Moab, Utah.

Razorback suckers have suffered a similar plight. Maligned by many who assign little or no value to them, suckers and their survival requirements were not well known for many years. Mink ranchers along the Gunnison River counted on spring runoff to bring a new crop of thousands of rich and meaty "humpback suckers" to feed their mink. Farmers along the Green River noted this strange sucker as it lay dying in large numbers in irrigation ditches and fields. Control of the river's flooding through dams and dikes hastened its demise. Without flooded bottomlands, there was no rich nursery habitat for the tiny young, and what few sheltered areas remained were filled with hungry non-native fishes, eagerly devouring the suspended slivers of flesh. Had it not been for their long life spans, the razorback sucker would now be extinct. But since individuals live for forty years, biologists have been able to effect recovery of this species through revitalization of flooded bottomlands, removal of non-native species, and reintroduction of hatchery-reared fish. The largest numbers of razorback suckers remain in Lake Mohave, trapped between Hoover and Davis dams in the lower basin, and in the Green and Yampa rivers of the upper basin. Still, biologists fear the species is losing genetic diversity and its ability to continue to cope in an altered environment.

While Colorado squawfish and razorback suckers are gone from much of their historic range, the humpback chub seems to have fared a little better. The species still remains in six of its original ten canyon locations. Ironically, its preference for canyon regions spelled its demise in some areas and its survival in others. Unable to survive in the artificial lake habitats and the cold water downstream of dams, humpback populations are gone from Flaming Gorge, Ladder Canyon, lower Cataract Canyon, and much of Grand Canyon. Populations at Black Rocks, Westwater, upper Cataract, Yampa, Desolation/Gray Canyons, and parts of Grand Canyon persist because these small areas happen to provide all the needs of the species' life history. The turbulent waters in these canyons restrict many non-native fishes from entering and thriving in the domain of the humpback chub. But reduced spring runoff has moderated the turbulent whitewater conditions and allowed a closely related and otherwise non-canyon species, the roundtail chub, to interbreed with

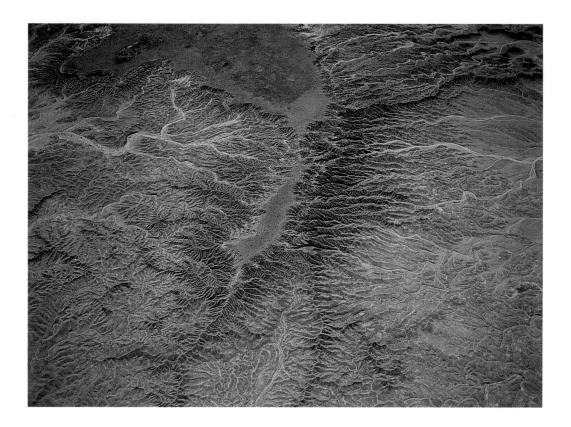

the humpback, possibly diluting the strong genetic bonds that have let both species survive for eons.

Hope for the survival of the bonytails seems bleak. As one of North America's most endangered fish species, wild bonytails remain in low numbers in Lake Mohave, and they are exceedingly rare in the rivers of the upper basin, where fewer than ten specimens have been found in the last fifteen years. Once reported from many regions of the basin through photographs and descriptions of its distinct slender tail trunk, the bonytails experienced a dramatic decline in the Green River following completion of Flaming Gorge Dam in 1963. Stocks held in hatcheries are progeny of a few wild adults, and loss of genetic diversity could already spell doom for this species.

The Colorado River will never be the river that the fish once knew. Their destiny has been fixed by recent human interference in a manner that has accelerated or superceded natural evolutionary processes. If the fishes of the Colorado River are to survive, we must make a balanced and rational commitment to manage the river in a manner consistent with survival of the fish while still meeting its needs for water delivery, flood control, and power generation. Large numbers of these fish can be raised in hatcheries. But these "aquarium" populations cannot be considered healthy, recovering species. Portions of the Colorado River can and must be managed in concert with human needs and still satisfy the life-history requirements of these fishes.

Facing page: Adeii Eechii Cliffs near Landmark Wash east of Cameron, Arizona.

Above: Chinle Formation of the Petrified Forest between Padre and Chinde Mesas, Arizona.

LARRY STEVENS DESCRIBES PLANTS AND A
FEW ANIMALS ALONG THE RIVERS AND
UPLANDS OF THE COLORADO RIVER BASIN.

A Telling Tree

Lawrence E. Stevens

*Facing page: Chinle Formation at the Adeii
Eechii Cliffs south of Tuba City, Arizona.*

Above: Barrel cactus.

AT GRANITE PARK in the lower Grand Canyon stands an old Goodding
willow tree with an incomprehensible history. It grows ten feet from the
river's bank. Woven into its roots and limbs are memories of the river in
flood and drought, of burning summers and freezing winters. Yet as in-
triguing as this single tree may be in its persistence, studying it as a distinct individual with a unique life history is not the usual approach of
science. Whether willow or swallow, an individual organism contributes
to the group of populations called a species, and it is to this collective
level that scientists usually turn their attention. Understanding this
Goodding willow, though, is a means to understand the daunting array
of ecological factors that determines the distribution of life in the Colorado River Basin, among them, temperature, light, soils, moisture, and
natural disturbances such as floods. Trying to unravel the story of any
single organism, such as this tree, is like trying to tease apart the strands
of a spider's web. But in doing so, we learn much about the riparian
ecosystem and come closer to understanding not only how that one
individual has succeeded, but how its individual success contributes to
the success of its species as a whole within the ecosystem.

For many living in or visiting the West, the word *riparian* tends to
conjure images of verdant fragile ribbons of vegetation bordering streams
and rivers, standing in sharp contrast against stark, arid, and often color-
ful landscapes. And though derived from the Latin *ripa* meaning bank or
shore, riparian as the term applies in the Colorado River Basin actually
describes a broad range of habitats, not just the biotic community abut-
ting rivers in this region. Wetland and riparian ecosystems occur in the
margins of ephemeral and perennial stream channels as well as in wet
meadows, and in areas of springs, seeps, hanging gardens, and marshes.
On the Colorado Plateau, riparian ecosystems, and the plant assemblages
and animal species that characterize and depend upon them, are not

Facing page, top: Little Colorado River at Dinnebito Wash downstream from Grand Falls, Arizona.

Facing page, bottom: Cottonwood along Pueblo Colorado Wash near Sunrise Springs, Arizona.

Above: Grand Falls of the Little Colorado River, Arizona.
Grand Falls was created when a lava flow plugged the Little Colorado River's course between Winslow and Cameron, Arizona. The river was dammed up behind the lava, but quickly flowed around its edge and back into its old channel. Part of the old channel, now downstream of the falls, was abandoned as the river hastily cut back upstream by headward erosion.

merely confined to the banks of rivers or to low elevations, but are found far into the uplands, even to tree line and tundra.

Riparian vegetation exists across a wide range of elevations, and in such divergent circumstances, its adaptation to varying water quality shouldn't be a surprise. The concentration of dissolved salt is generally low in headwater reaches, but as river water moves downstream, it usually accumulates salt from saline springs and local runoff. Springs with a low-salt concentration, such as Thunder River in Grand Canyon, support lush stands of deciduous vegetation. Saline springs such as those emanating from Grand Canyon's Tapeats Sandstone, may only be colonized by halophilic (salt-loving) plants, such as the non-native tamarisk. In its headwaters in Arizona's White Mountains, the Little Colorado River is a clear, cold-water mountain stream, supporting the highest diversity of willow species on the southern Colorado Plateau. But the Little Colorado loses all of its headwater flow near Holbrook, Arizona. Here it

Above: Flashflood in Landmark Wash flowing into the Little Colorado River east of Cameron, Arizona.

Right: Fields near Willow Springs west of Tuba City, Arizona.

becomes a salty ephemeral stream dominated by tamarisk and camelthorn, with a few large, old native Fremont cottonwoods. Near its confluence with the mainstream Colorado River, Blue Springs adds two hundred cubic feet per second of calcium-rich water, and makes the river far too saline to support all but the most halophilic plants: tamarisk, camelthorn, arrowweed, saltgrass, and common reed.

Plants occupying wetland and riparian habits all share an adaptation enabling them to survive water-logging. Called phreatophytes, the roots of such plants are in contact with the water table or its overlying capillary fringe. Common phreatophytic plants, such as willows, cottonwoods, alders and water birch, are capable of withstanding prolonged inundation during flooding. A relatively small proportion of the basin's higher plant species are hydrophytes, fully aquatic species that can live under complete or prolonged submersion.

High-elevation wetland and riparian habitats occur widely in the Rocky Mountains, the Uinta Mountains, and in six isolated mountain ranges on the Colorado Plateau—the La Sals, Abajos, and Navajo Mountain east of the Colorado River; the Henry Mountains west of the river; and the San Francisco and White Mountains south of the river. The highest elevation plant assemblages occur in tundra ecosystems around twelve thousand feet on the peripheries of permanent snow fields. There, dripping water during the summer months supports dense mats of sedges and, where space is available, grasses, rushes, and an extraordinary array of wildflowers. This environment is fraught with physical stresses, particularly freezing, frost heaving, lightning strikes, and heavy wind, all of which force plants to grow rapidly. Many alpine plants, for example, bloom three weeks after snow melt and produce seed three to five weeks later. Soil distribution, the scouring of wind, the timing of snow melt and the availability of water within the soil play important roles in alpine plant community development. As a result, alpine vegetation is a mosaic of plant associations.

Alpine wet meadows, such as those in the upper Green River drainage, occur at altitudes of eight to ten thousand feet above sea level, and are dominated by low-growing plants such as Bebb's willow and a number of rushes and sedges. The Uinta Mountains host a large number of alpine plant species. Slightly below tree line, three types of subalpine riparian meadows can be found on the White River Plateau between ten and eleven thousand feet: tall forb, short forb, and Thurber fescue grass meadows. Slightly lower yet, tributaries of the Strawberry River in Utah are dominated by aspen.

Below elevations of six thousand feet, numerous shrubs and several trees, such as narrowleaf cottonwood and box elder, dominate the river banks. The Yampa River in Colorado offers a characteristic example of riparian vegetation at these altitudes, descending through coniferous forests to pinyon-juniper, sagebrush, and mountain scrub vegetation before it joins the Green River in Dinosaur National Monument in Utah. Downstream, the lower Green River flows by Fremont cottonwood, skunkbush, non-native tamarisk, and other opportunistic upland species. After joining the upper Colorado River and passing through Cataract Canyon and Lake Powell, the Colorado River's natural streamside

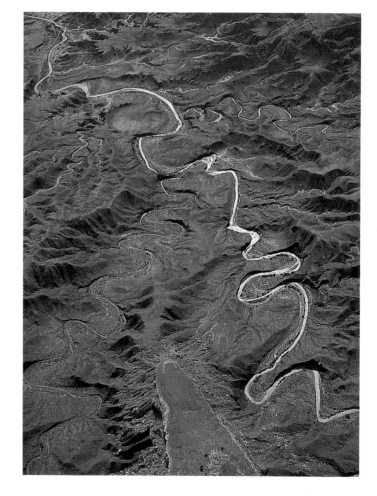

Above: Hah Ho No Gah Canyon east of Tuba City, Arizona.

Right: Moenkopi Wash northeast of Tuba City, Arizona.

Above: Autumn amidst the spruce and aspen on the northwest side of the San Francisco Peaks, Arizona.

vegetation in wide reaches consists of Sonoran Desert Goodding willow, mesquite, catclaw, graythorn, and desert broom. The narrow reaches of the river in Grand Canyon have relatively little natural riparian vegetation.

Seeps, springs, and hanging gardens are particularly memorable for visitors to the Colorado Basin—perhaps because they are unexpected and consequently all the more delightful. Groundwater, silently flowing beneath the surface in such places provides essential support for many unique riparian species.

Although they are small, seeps are numerous and support endemic plant species such as McDougall's flaveria and Navajo sedge. Springs often emerge along fractures and faults which serve as conduits for groundwater flow. Some of the most pristine springs emerge from the base of the Redwall Limestone in the Grand Canyon, including Vaseys Paradise, Cheyava Falls, and Thunder River. Profuse stands of crimson monkeyflower, cardinal flower, and redbud, as well as willow, cottonwood, and ash trees border these natural oases.

At hanging gardens, such as Weeping Rock in Zion National Park, a spring emerges from rock near the base of an aquifer. Freeze-thaw cycles in these wet areas may erode the cliff face into an overhang, creating a shaded, dripping back wall. These back-wall habitats are often colonized by maidenhair fern, and immediate surrounding areas are occupied by lush beds of orchids and other rare species, surrounded by a larger zone of perennial shrubs, such as Utah serviceberry and silk tassel, and trees such as Fremont cottonwood, hackberry, ash, and scrub or Gambel oak.

Habitats at springs are relatively small, isolated, and are used opportunistically by large animals; however, they often support endemic invertebrates and occasionally amphibians, including rare and endangered species. The Kanab ambersnail is presently known at only two springs on the Colorado Plateau. It was first discovered near Kanab, Utah, and

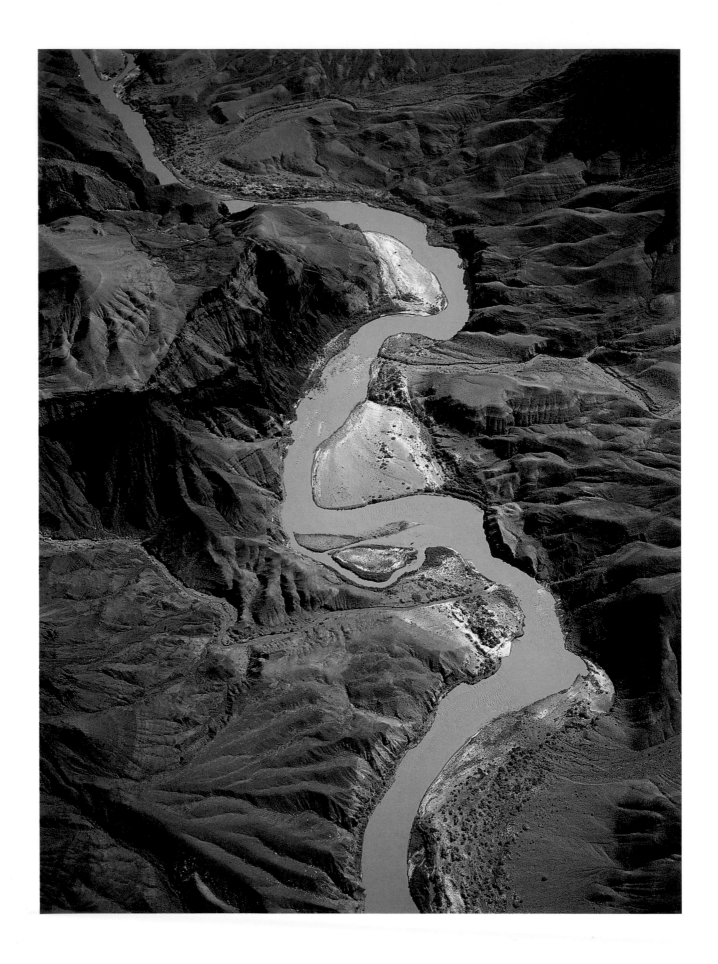

Facing page: Colorado River at Tanner Wash in Grand Canyon, Arizona.
When Major John Wesley Powell first floated through Grand Canyon, he saw a river dominated by sand beaches. Deprived of sand by Glen Canyon Dam, many of the beaches now have eroded to small vestiges of those seen by Powell a century earlier. The Paria and Little Colorado Rivers still deliver sediment to the Colorado that can be reworked into beaches by intermittent floods. Sandbars such as these near Tanner Wash are maintained when the river is allowed to flood.

Above: Crossbedded Navajo Sandstone, Zion National Park, Utah.

the largest of the two remaining populations occupies a cattail/sedge/willow-dominated wet meadow at Vaseys Paradise in Grand Canyon. Similarly, individual populations of the rotund crawling waterbug occupy dripping mossy springs at low elevations on the southern Colorado Plateau. The only other populations known are found in similar habitats in the mountains of southern Mexico. At its lowest elevational limits, the northern leopard frog exists primarily at springs and marshes, and a population was recently discovered at a spring just upstream from Lees Ferry near Glen Canyon Dam.

When we encounter the dramatic landscapes of the Colorado River Basin, it's difficult for human beings with limited life spans to imagine the astounding changes this landscape has undergone over the vast expanse of geologic time. Oceans, bays, river deltas, and deserts are but a few of the many environments that have existed there. Each native plant species in the present riparian ecosystem has evolved under its own unique combination of these environmental and biological conditions, and has survived enormous environmental change. The horsetail lineage, for example, can be traced back through the fossil record more than 350 million years. Numerous fern fossils in the Hermit Shale in Grand Canyon

Above: Virgin River tributary between Mesquite and Overton, Nevada.

Right: Iceberg Ridge, Lake Mead National Recreation Area, Arizona/Nevada.

indicate that a profuse terrestrial flora grew in the region during the Permian Period, 250 million years ago. Near Westwater Canyon on the Colorado/Utah border, the seventy-million-year-old Dakota Formation contains at least nineteen species of fossil ferns, five species of sycamores, and one species of willow. Nearby, located along what was then an ocean coastline, were tropical and subtropical plants, while in the highlands early conifers could be found.

Mountain-building geologic events in the Cenozoic era gradually created the rain shadow climate that affects much of the Colorado River Basin. Fifteen million years ago, the middle basin supported oak, chaparral, and grasslands, with forests to the north and shrublands to the south. The basin dried out during the Pliocene Period, about five million years ago, and water-loving plants species became increasingly restricted to wetland habitats. Therefore, many riparian plant species are derived from lineages as ancient as some of the rocks of the Colorado Plateau itself.

Understanding that the Colorado River Basin has a history predating ours by millions of years might lead to the falsely comforting idea that human beings can't possibly impact or harm such a venerable, durable landscape and its ecosystems. In fact, human activities have dramatically altered riparian habitats in a number of ways.

Hundreds of dams, diversions, and other modifications of the streams and rivers have altered both ground and surface water supplies, affecting every element of the riparian ecosystems throughout the Colorado River Basin. Large reservoirs with immense shorelines have been created in places where natural ponds and lakes would be relatively rare. Since Glen Canyon Dam was completed in 1963, lower riparian zone plants, particularly coyote willow, horsetail, seep willow, and non-native tamarisk, camelthorn, and the brome and love grasses, have extensively colonized the Grand Canyon downstream. Post-dam riverside marshes have become relatively abundant, a novel development that indicates increased ecological stability and productivity. In response to this increased riparian vegetation, there is now a higher diversity and production of invertebrates, Neotropical migrant birds, including the endangered Southwestern willow flycatcher, and many other terrestrial vertebrates.

But the pocket of higher biodiversity and productivity in Grand Canyon is exceptional. Human activities have eliminated at least thirty percent of the Southwest's wetland and riparian habitat over the past two hundred years, through mining, grazing, recreation, and the introduction of non-native species of plants and animals. This percentage of loss is even more significant because, in terms of biodiversity, riparian habitats are disproportionately important landscape elements. In the Colorado River Basin, these habitats comprise less than one percent to more than fifty percent of the landscape area (from low to high elevations, respectively). Because the paucity of water is the defining characteristic of this region, riparian ecosystems, where water is concentrated, are the most biologically productive and diverse habitats.

*

John Wesley Powell, as the second Director of the U.S. Geological Survey, envisioned a concerted, basin-wide approach to the management

Above: Bill Williams River below Alamo Crossing, Arizona.

Left: I-40 and Santa Fe Railroad crossing Colorado River above Topock Gorge, Arizona/California.
The Arizona and California shores of the Colorado River are here stitched together by transcontinental pipelines, highways, and a railroad. The Havasu National Wildlife Refuge lies above Topock Gorge in a marsh that the river once meandered across before being channelized.

Above: Farming near Poston, Arizona, along the Colorado River.
Surprisingly little farming is done along the banks of the Colorado and Green Rivers beyond their upper reaches in Colorado and Wyoming. Irrigated fields do reappear briefly in the vicinity of Parker, Arizona, and Blythe, California.

of the plateau's water and land resources. Today, several Department of the Interior agencies and Native American tribes share most of the responsibility for managing the region's rivers, including the National Park Service, the Bureau of Reclamation, and the Navajo and Hualapai tribes; however, at present, administrative coordination between the various river segments is limited.

In large measure, the seeds of Powell's vision fell on dry ground. Basin management has too often been piecemeal, laid out along political rather than natural ecological boundaries. Now, in addition to managing the human-imposed changes on wetland and riparian habitats in the Colorado River Basin, we have the added challenge of trying to preserve what remains of them.

Although the lone remaining Goodding willow at Granite Park offers a testament of resilience and endurance, all is not well in the Colorado River Basin's riparian ecosystems. There may be little or no possibility of reversing some of the harmful changes we've caused. The bleak legacy of decisions made in recent decades may be part of a longer, even bleaker bequest we make to our descendants. But a fundamental question remains: at what level shall we balance long-term economic and environmental sustainability on the Colorado Plateau and in the world in general? The answer to this uniquely human question is urgently needed.

Water, Earth, and Sky: The Colorado River Basin

ELLEN MELOY EXPLORES HER PERSONAL RELATIONSHIP WITH THE RIVERS OF THE COLORADO BASIN.

The Silk that Hurls Us Down Its Spine

Ellen Meloy

Facing page: Colorado River below Ehrenberg, Arizona.

Above: Evening primrose.

IMAGINE THE COLORADO RIVER BASIN as if it were a grand banquet spread before you, your dining room chair bellied up to the Grand Canyon, your left elbow resting on Hoover Dam, your right elbow on the Sleeping Ute's nose, your fork poised somewhere above Canyonlands National Park.

Think not of a tectonic plate but of a sumptuous feast. Mesa tops of thick-headed pinyon-juniper broccoli, meandering banks of lush cottonwood celery and tamarisk slaw, a tangy salad of hackberry, coyote willow, and other riparian greens. Rich, teeming eddies of catfish bouillabaisse and carp carpaccio. Slickrock pools of quivering green Jell-O, sage-freckled Uinta Basin custard, Book Cliffs tortillas, frybread rolled from the yeasty mounds of Nokaito Bench. The cool slake of a pothole martini, garnished with tadpole shrimp and a Russian olive. Wingate Sandstone tarts steaming beneath a latticed cryptogamic crust. North Rim cutlets breaded in mountain bikers. A jumbo helping of Moenkopi mud pies piled high on your sectional plate, buttressed with brave volcanic dikes to dam off the gravy of Chinle Wash. A thin ribbon of cafe au lait river that carries a tiny speck of a human being, although the lusty feast distracts your notice.

You gulp succulent brachiopods embedded in limestone fruitcakes, gnaw the bony ribs of Shiprock, spoon up sun-ripened tomato-Red House Cliffs, and ice your tongue with the San Francisco Peaks' pale sherbet. You swallow Navajo Mountain like a plump muffin and bite off that potato chip of a dam on Glen Canyon. A faint tease of gluttony numbs your palate. Or is it indigestion? Still, you must eat, for each year the Colorado Plateau menu diminishes, the diners grow more numerous and their appetites, ravenous.

Munch that meatloaf of a mesa, that tenderest loin of Comb Ridge. Wash down Monument Valley with foamy drafts of Lava Falls. Crunch

Above: Los Angeles Aqueduct drawing water from Lake Havasu and Colorado River above Parker Dam.

Right: Diversion dam on the Colorado River north of Ehrenberg, Arizona, and Blythe, California.

those toad bones, devour the lizards, ravage the Triassic cephalopods. Never mind the desert bighorns, who are endangered, or the ravens, who taste awful. Avoid the datura. Spit out the dung beetle. Relish bite-size canyon wren appetizers, impaled on yucca toothpicks. Slurp the dark sauces of Black Mesa coal, fend off scurvy with Lime Creek, and scald your tongue on Mexican Hat's gravel-capped uranium tailings pile, that broad, glow-in-the-dark, blue cornmeal-mush wedge sprawled below and beyond the Moki Dugway—insane and forever.

After-dinner Cigarette Spring in hand, plate clean clear down to bedrock, you push back your chair and loosen your belt. You pluck a few B-52s from your teeth and launch a sated burp toward Denver. If we gourmands truly love this river basin, you sigh, we shall never go hungry.

Life as a desert omnivore, however, is never simple. The creeping heartburn returns. Underneath the plateau's feral pelt, some pathetic, squiggling thing on the tablecloth, some fly in the river soup, catches your eye. Go back to that canyon-bound ribbon of water with the little *Homo* in it, the one swimming the river like a desperate bug.

*

The trouble seems to be this: although I should be in my raft, at the oars, savoring a morsel of uncooked Utah, I am not in the raft. On hundreds of trips down Colorado Plateau rivers over a span of eighteen years, I have been in the water on countless voluntary swims, but I should not be in the water now. Although I have often been stupid, I have just been extremely stupid. My need to pursue absurdity has overwhelmed my need to stay out of trouble. Moments ago I rowed to shore to investigate a pair of bright yellow objects in a debris pile. I pulled the raft's nose up on the sand and clipped my lifejacket to the frame. I walked away. While my back was turned the river nibbled the raft off the sandbar and carried it off in its swift current. The raft is empty. I am swimming after it.

The boat holds lifejacket, transportation out of the desert, drinking water, and food—real food, not some carbo-loaded physiographic province. The ghost voices of everyone who taught me anything about river running have stopped screaming *Always tie up your boat* long enough to start screaming *Always stay with your boat.* I am staying with my boat—it happens to be a considerable distance ahead of me, careening down the canyon. I am alone. I might drown. I have not finished living yet.

This must be the moment for a split-second flashback of my entire life—learning to read, eating avocados, my hamster Ned, my flannel sleeping bag with the little green turtles, climbing Mt. Whitney, falling in love. However, I shall go back only as far as morning.

*

The coyotes howled the sun up the canyon rim and when it spilled into my camp, the songbirds surrendered such lovely notes, they broke my heart. So fervently did they sing, it was impossible not to separate their songs from all the other sounds of daybreak and fix on them as I lay wide-eyed and broken-hearted in my burning bed.

The canyon already blazed with heat. Sleeping bag and clothing felt like tools of the Inquisition. I donned hat and sunscreen and flicked aside the bed tarp quickly, checking for scorpions. On the camp stove I brewed coffee then cooled it down and iced it, barely surviving this rigor of Life

108

Facing page, top: Colorado River in the Imperial National Wildlife Refuge north of Yuma, Arizona.
The Imperial National Wildlife Refuge is now located on land that was once beneath the waters of the Imperial Reservoir. But the Colorado quickly filled the lake with sand, and it again flows as a river almost to the foot of the dam.

Facing page, bottom: Colorado River through Imperial National Wildlife Refuge, Arizona/ California.

Above: Eagle Creek above its confluence with the Gila River, Arizona.

in the Wilderness. Major John Wesley Powell, who ran the Green and Colorado rivers more than a century ago, never had block ice in a 120-gallon Gott cooler lashed to his wimpy wooden skiff. While most of today's river runners would gag on the Powell diet of rancid bacon, moldy flour, and stale coffee, they would sell their mothers for a glimpse of the Colorado Plateau before the dams and the crowds. I bore no delusions that the river beside me still flowed through Powell's *terra incognita*. From Rocky Mountain headwaters to the Sea of Cortez, the Colorado shoulders the weight of our needs, the thirsts of our farms and cities, and the affection of thousands of recreationists who ply its waters.

I daydreamed as I broke camp and loaded the raft, a routine repeated many times in my gypsy life as a river ranger's wife who spends much of the season afloat with him on his patrols through the canyon. The packing, hefting, loading, and strapping felt so reflexive that, even after the first night out, I forgot that I had left the ranger behind. I missed him but

Above: San Francisco River upstream from Clifton, Arizona. The Gila River, fed by the San Francisco River, Eagle Creek, and other tributaries within southwestern New Mexico and southeastern Arizona, is a free-flowing river through much of its course. Periodic large floods still maintain an open channel on this river until it is bridled by San Carlos Dam south of Globe, Arizona.

Left: Dry wash leading to Gila River below Redrock, New Mexico.

Above: Gila River bed above Painted Rock
Reservoir, Arizona.
The Gila River is a thicket of mesquite and
tamarisk as it rolls through its Great Bend
southwest of Phoenix, Arizona. Upstream dams
and water withdrawals have prevented floods
that would have cleared out this vegetation and
maintained a more open channel.

I did not need him. Other than keeping my tasks well within my own strength, a solo trip did not seem like a big deal.

I untied the raft, pushed it into the water, and climbed aboard. After an easy ferry the raft slipped into the main current. With an abruptness that could detach your teeth from your head, a low-flying B-52 suddenly unzipped the air between canyon rims, followed by an ear-splitting roar. Military planes flew over the river daily, sleek and darkly Transylvanian, looking as if no one was in them but bombs and robots. Around the first bend I passed a party of river runners. They waved from their camp, obviously bearing up well under the stress of the Nuevo Powell style—rancid prosciutto, stale cappuccino, Utah beer. Then I had the river to myself. It belonged to me all day long.

The canyon walls rose high, voluptuous, and red. I felt the river's muscle beneath the raft, pulling me along with little effort from the oars. It came to mind how simple river life is: surrender all human measure of

the lurchings of the universe to the river's own gauge of time and distance. Shed reason and frets so that what is left is a lean asceticism, a looking not at the world but into it. Let the morning serve up, as every slice of desert river will, a great blue heron.

The morning's heron stood in the shallows ahead, toes planted like twigs in the sandy bottom. It stared its Pleistocene stare. One foot lifted so slowly, the droplets that fell from it barely disturbed the water's surface. In a lightning motion the heron's stare disappeared underwater then reemerged. A few rope-like jerks of the bird's neck sent a silvery fish down its gullet. You can never get close enough to hear a heron's throat, nor can you easily identify the meal before it slides down that hose of a neck. Exotic and native fish swim these silty waters, natives whose endemism to the Colorado River Basin is ancient by several million years and now precarious. Biologists call the endangered Colorado squawfish, humpback and bonytail chubs, and other natives *obligate species*, creatures limited to a certain condition of life—that is, this river in which they evolved. In an ecology dramatically altered by humankind, some have become extinct, others hang on by a fin.

Unless you've been dead you know that this is a land replete with consequence and complications. We know our own history: twentieth-century humans came to this arid, fissured, difficult place blinded, seeing not its particularities but our own aims, and because of this blindness, the shackles were severe, our covenant with nature set in terms of conquest rather than compatibility. "Taming the menace," in the words of the U.S. Bureau of Reclamation, has transformed the Colorado into a push-button river with a queue of dams and diversions in the lower basin and a flood-poor river in the middle basin. We reap the bounties of a watershed plumbed from headwaters to mouth, and we know the environmental costs—among them, the loss of native fish, the drowning of spectacular canyons under enormous reservoirs, the alteration of riparian communities. We know that our urban desert culture is quickly outstripping the Colorado's ability to support it. What we do not know is, have we learned anything? Would we do it all over again?

In hundreds of side canyons and washes, on semiwild pieces of mainstem bracketed by reclamation hardware, fluid gravity still more or less makes its own decisions and follows the path of its own weight, taking with it worn-down pieces of the continent grain by grain. It is in these reaches that you are most likely to glimpse the river's fight for eternity. To witness that persistence is why many of us make our home here. Writer Richard Manning said, "A river cannot be described, or rebuilt from its elemental facts, just as it cannot be imagined alive into images. It must be lived." For me the bond between self and place is not conscious—no truth will arrive that way—but entirely sensory. Instinct and intimacy bring the feast closer, the river celebrates things we forget how to celebrate: our own spirits, the eternity of all things.

Watching the heron brought to mind a birdwatcher's story from the much-reengineered lower Colorado, where a meandering swath of sand and vegetation overgrew an old, dewatered channel. Instead of flying in a straight line across the flats, a great blue heron followed the dry serpentine curve of the ghost channel, unable—or unwilling—to disobey an

Facing page: Picacho Peak in California, north of Yuma, Arizona.

Above: Sand and gravel operation on Salt River east of Phoenix, Arizona.
The Salt River once carried steamers past a tiny village called Phoenix. Now its flow is held behind four dams in the mountains to the east, released for consumption by the two and a half million people who live in the Valley of the Sun. The river still occasionally floods, spilling into the Gila River downstream from Phoenix.

Right: Gila River near Growler, Arizona.

Above: Sonoran desert around Vekol Wash,
west of Casa Grande, Arizona.
Desert washes are dry more often than not, but
their channels are still determined by the
variables of flow and flood that guide rivers in
a wetter terrain.

ancient avian wisdom, the imperative to match its flight to the shape of a river.

As the raft approached, the morning's heron took wing and flew the river bends as I did. The bird paid no attention to the unlikely yellow objects set against a gray-brown stack of logs jammed against the bank. The attention was mine. I pulled on the oars to make shore.

*

I could use about twenty herons now, herons to pluck me from this roiling sandstone gravy and drop me into the runaway raft like a wet noodle. The water is not cold but the flow is swift—no rapids yet, few rocks to bash myself against. In the shallows, without a lifejacket for buoyancy and padding, the river bottom could shred me like a cheese grater. I struggle to stay in the deepest current, to reduce the distance to the boat, but my legs move like Gumby limbs, too weary to kick much longer.

What happens when I surrender to this aloof, silken creature that hurls me down its spine? What happens when I exhaust my strength and can swim no more, no matter how hard I try or how desperate I am? What happens if I let go?

How very familiar these waters feel, waters brought in from all directions over broad stretches of desert, through sinuous canyons to join their downward run to the sea. Every leaf, every riffle, every russet sweep of sandstone is saturated with memories, cut through and through with my own history. Along the river are strewn stories that make me who I am, that tell me what binds my life together, what to cherish and what to lay aside. We know few ways to tell the river's story other than through our own. Yet here it flows, making beauty a fact of existence, visible in its

Facing page, top: Fields along the Gila River at Ligurta, Arizona.
Floodways from the surrounding mountains deliver occasional torrents directly to the Gila River that would otherwise inundate fields and make farming more difficult.

Facing page, bottom: All-American Canal taking Colorado River water through Algodones Dunes west of Yuma, Arizona.
The All-American Canal delivers water to California's Imperial Valley, one of the most fertile and productive agricultural centers in the United States. In 1906, the Colorado River broke out of its established channel and drowned much of the Imperial Valley, creating the Salton Sea.

Above: Dry wash channelized over the All-American Canal north of Yuma, Arizona.

presence and absence, in flood and in washes as parched as old bones. To witness it you are grateful to have senses, to feel this one exquisite and dangerous thing that holds you to all life.

Shadows carve bands of mauve into the sun-bright terra-cotta of the canyon walls. Ripples of heat quiver against the highest rims and disappear into an azure vault of sky. Ahead the river's surface splinters into a million fragments of sunlight. I try to match my swim to the shape of the river.

The river carries the raft into an eddy and slows it down. The oars no longer spin wildly in the oarlocks. If you observe stream hydraulics carefully you will know that most eddies move in circles. The main current meets the eddy's slower water and pushes it in the opposite direction— the river flows upstream. It curls back then meets the current again. At the top of the circling eddy I tread water and wait for the river to deliver the raft to me.

Never have I been so happy to become one with a gray bubble of inflated Hypalon. From water level the pontoon looks as high as a blimp. I find a strap end and use it to haul myself up and into the boat. Like bedraggled kelp, my dripping body drapes over the fat tube. A sudden anger rolls red-hot bowling balls through my skull, making ungodly, entirely negative noises. I kick the sides of the dry box, the oars, the aluminum frame. I kiss my lifejacket. I bite the bow line to shreds. I squint and growl, *Stupid, stupid, stupid.* All around me lies the sumptuous Colorado Plateau feast. In the cooler lies a normal dinner. In the pocket of my soaked shorts lie items scavenged from a debris heap. Bathtub toys. Two yellow plastic ducklings.

*

In the last hour of sun the river that ran jade during the day runs with flames of copper and indigo. The shadow cast by a canyon wall already holds dusk, its edge a sharp clarity. Bathed in the apricot light of

day's end I watch the dark shadow move across the water. Oddly, the serenity of being beside a river induces an intensity of awareness that is inflammatory—one moves from languid calm to burning ecstasy in an instant. Inside the Jurassic walls, my frail body nearly explodes with the sheer beauty of this place. It is as if there is no air or stone here, only light.

I have guarded against theft by nearly dry-docking the raft on the beach and tying both bow and stern lines with an overkill of knots. I arranged tarp, bed, and ducklings under a juniper, intending to inhale the tree's scent all night, and I ate a monkish repast. I feel no loneliness, only aloneness. Do I like being where I am, by myself? I do. I can sit on the bank and stare. If I want to, I can drool. I can absorb without distraction the bursting fullness of light and living things that give the inert sandstone a tangible pulse. In solitude you strip yourself bare, you rest your mind on what is essential and true. The canyon's shadow engulfs camp. For a while tendrils of warm air from the day-baked cliffs stave off the shadow's cool.

The river's riches, like life's memorable moments, often assume a dreamlike quality when you think back on them. Although it felt like forever, I had not actually chased the raft very far; it was not a near-death experience. Did this harsh, red-boned desert impart something besides its indifference to my wholly insignificant, lifejacket-less hide? Have I learned anything?

I have learned that, skill notwithstanding, the river is always in charge. It can mesmerize and tranquilize, pulverize and tenderize, rip your limbs apart, fill your lungs with mud. Canyon winds can turn your boat upside down on the water, the July sun can turn your brain to mush, ice in the eddies can lacerate your hands with its shards. The river has taught me that thunder cues toad lust, and a lizard I once watched turned its belly cobalt blue then de-blued it before my very eyes. That ravens don't like dirty toes, so they clean them often. That the Anasazi, prehistoric desert farmers who scattered their petroglyphs throughout the canyonlands, were as interested in human genitalia as we are.

I have learned that when we dammed Glen Canyon we traded a stunning redrock wilderness for better hospitals, security lighting in Kmart parking lots, lettuce in January, and block ice in my boat cooler. That beneath the slackwater reservoirs flows an outlaw river still intending to reach the Gulf of California. That a chunk of Entrada Sandstone the size of an obese Honda can drop out of the sky and so, too, can fall the tiniest bird's egg with a shell the color of air and a yolk as red as blood. That my husband's devotion remains steady and true, to me and to the river, and that after hiking a sidecanyon in 103-degree heat, a cantaloupe from the cooler was the best thing we'd ever eaten in our entire lives. I have learned that on the clearest days in the worst drought, cottonwood leaves sound like rain when they tremble, and if you walk far enough you can still find prayer bundles tied to the branches of coyote willows beside a spring that gushes out of bare rock with such power it seems possessed.

I have learned that when I lie on a sweep of polished slickrock in a certain canyon near Navajo Mountain, my body and the desert are the same shape, a perfect fit of rock and flesh. The river draws off my mad-

Facing page, top: Desalination plant on the Colorado River below Yuma, Arizona.
The desalination plant was built in an attempt to deliver less-saline water to Mexico. Failing to live up to its technological promise, the plant is currently mothballed. The Colorado River trickles past in the background, heading for Morales Dam and Mexico.

Facing page, bottom: Dry bed of the Colorado River near Gadsen, Arizona.
What little water remains in the Colorado River as it enters Mexico is taken out at Morales Dam and distributed to fields in Sonora and Baja California.

Above: Colorado River and the Sea of Cortez, Sonora/Baja California, Mexico.
The Colorado River now rarely reaches the Sea of Cortez. Its estuary is washed by tidal flows.

Above: Tidal flats, mouth of Colorado River at the Sea of Cortez, Sonora/Baja California, Mexico.
The Colorado River's delta is beveled down by tides that carve feathery channels as they withdraw back to the sea.

ness and calms me, it knows nothing of my love for it, it can, like love and mystery, prove unattainable even in the moments of profoundest intimacy. The river flows downstream. Sometimes it flows upstream. The physics of river can show you endless facets of gravity made visible and fluid. It can carry two yellow ducks from God-knows-where or for how long or how many miles, and set them down in the same debris pile, three inches apart.

The canyon takes on moonlight long before the moon itself becomes visible. The delirious bird opera that opened the day has given over to a soothing whisper of water flowing by, never stopping. The night songs come from a poorwill. Filled with formless, silvery light, the gorge floats as if weightless then comes into sharp focus when the luminous orb tops the rims. There is no end to appetite here. This lusty feast of a province may be better suited to nourish our souls than our material excesses. We eat it to build cells and sinew, to run through our own veins and resupply our blood.

The moon shivers in the small ring of dark inside my teacup. On my two-thousandth night on the river, I still don't know anything. The best I can do is to always tie up my raft when I leave it. The best I can do is follow the serpentine curves, the dry, ghost river as well as the wet, flowing one. I am an obligate species, obligated to have this river. I can match this life to its shape. Perhaps then I might learn something.

SELECTED READINGS

Carothers, Steven W., and Bryan T. Brown. *The Colorado River Through Grand Canyon.* Tucson: University of Arizona Press, 1991.

Collier, Michael, Robert Webb, and Edmund Andrews. "Experimental Flooding in Grand Canyon." *Scientific American* 276, no. 1 (1997): 66–73.

Collier, Michael, Robert Webb, and John Schmidt. *Dams and Rivers, Primer on the Downstream Effects of Dams.* U.S. Geological Survey Circular 1126 (1996).

Harper, K. T., L. L. St. Clair, K. H. Thorne, and W. M. Hess. *Natural History of the Colorado Plateau and Great Basin.* Boulder: University of Colorado Press, 1996.

Hundley, Norris. *Water and the West—The Colorado River Compact and the Politics of Water in the American West.* Berkeley: University of California Press, 1975.

Meloy, Ellen. *Raven's Exile: A Season on the Green River.* New York: Henry Holt and Company, 1994.

Minckley, W. L., and James E. Deacon, eds. *Battle Against Extinction.* Tucson: University of Arizona Press, 1996.

National Resources Council. *Colorado River Ecology and Dam Management.* Washington, D.C.: National Academy Press, 1991.

Webb, Robert H. *Grand Canyon, a Century of Change.* Tucson: University of Arizona Press, 1996.

Facing page: San Juan River at Comb Wash, downstream from Bluff, Utah.

Above: Box elder leaves.

CONTRIBUTORS

E. D. ANDREWS is Chief of the U.S. Geological Survey's River Mechanics Project based in Boulder, Colorado. He earned a B.S. at Stanford and his Ph.D. in geophysics at University of California at Berkeley. He rowed for Grand Canyon Dories from 1969 through 1974 before joining the survey. His most satisfying work has been helping to design the Grand Canyon experimental flood of 1996 and establishing water rights for instream flows at Zion National Park.

MICHAEL COLLIER is a family physician in private practice in Flagstaff, Arizona, and works as a writer/geologist for the U.S. Geological Survey out of the Desert Research Laboratory in Tucson. He received his B.S. from Northern Arizona University and M.S. from Stanford, both in geology, and his M.D. from the University of Arizona. He rowed the Colorado River commercially during the 1970s and early 1980s. He has written and photographed books on the geology of Grand Canyon, Capitol Reef, Denali, and Death Valley. His most recent work is on the San Andreas Fault in California.

ELLEN MELOY is the author of *The Last Cheater's Waltz* and *Raven's Exile: A Season on the Green River* (Holt), winner of the Spur Award for contemporary nonfiction. The Whiting Foundation honored her with a Writer's Award in 1997. Her naturalist essays have been widely anthologized, and she has written for *Northern Lights, Orion,* and other journals. Meloy lives on the San Juan River in southern Utah.

JOHN C. SCHMIDT received his Ph.D. from Johns Hopkins University and is now an associate professor in the Department of Geography and Earth Resources at Utah State University. He has conducted research on the Colorado River in Grand Canyon since 1985 and on the Green River in Utah since 1991. He is especially interested in the effects of climate

Facing page: Paria River gorge above Lees Ferry, Arizona.

Above: Fremont petroglyph.

Above: Colorado River at 36-Mile Rapid in Marble Canyon, Arizona.

and dam operations on channels, and in understanding the relationships between the physical habitat of the Colorado River system and its native fish.

LAWRENCE E. STEVENS was born in Cleveland, Ohio. He received his undergraduate degree from Prescott College and his M.S. and Ph.D. from Northern Arizona University in Flagstaff. Stevens has served as an ecologist to the National Park Service and the Department of Interior's Grand Canyon Monitoring and Research Center. He is an adjunct faculty member in the Department of Biological Sciences at Northern Arizona University, and a research associate at the Museum of Northern Arizona. He is an avid natural historian and river runner, and he has spent the last quarter-century engaged in ecological research. His studies have focused on native and non-native vegetation dynamics, as well as ecological processes in the Colorado River ecosystem. He presently lives in Flagstaff with his daughter Phoebe.

RICHARD A. VALDEZ is Senior Aquatic Ecologist for SWCA, Inc., and has worked for the New Mexico Game and Fish Department and for the U.S. Fish and Wildlife Service in the Colorado River, the Great Basin, and in Yellowstone National Park. He is a certified fisheries scientist with the American Fisheries Society and is a member of the Colorado River Fishes Recovery Team. He conducted many of the early surveys and life history studies of fishes of the Colorado River, particularly in whitewater regions, including Cataract Canyon, Westwater Canyon, Desolation Canyon, Black Rocks, and Grand Canyon. He has published more than thirty papers on the ecology and conservation of western fishes.

DAVID L. WEGNER is an ecologist who spent twenty-two years with the U.S. Bureau of Reclamation. From 1982 to 1996 he headed the bureau's Glen Canyon Environmental Studies unit. The findings of that project led to the Grand Canyon experimental flood of 1996. He is currently serving as vice-president of the Glen Canyon Institute.

Following page: Tidal flats, mouth of Colorado River at the Sea of Cortez, Sonora/Baja California, Mexico.

Book design by Richard Firmage, Salt Lake City

Typeset in Adobe Minion and Adobe Hiroshige typefaces
and composed by the designer.

Printed on 128 gsm Japanese Matte Art paper.

Printing, color separations, and binding by
C & C Offset Printing Company, Hong Kong.